Haptics Technologies

Springer Series on Touch and Haptic Systems

Series Editors

Manuel Ferre
Marc O. Ernst
Alan Wing

Series Editorial Board

Carlo A. Avizzano
José M. Azorín
Soledad Ballesteros
Massimo Bergamasco
Antonio Bicchi
Martin Buss
Jan van Erp
Matthias Harders
William S. Harwin
Vincent Hayward
Juan M. Ibarra
Astrid Kappers
Abderrahmane Kheddar
Chris McManus
Miguel A. Otaduy
Angelika Peer
Trudy Pelton
Jerome Perret
Jean-Louis Thonnard

For other titles published in this series, go to
www.springer.com/series/8786

Abdulmotaleb El Saddik
Mauricio Orozco
Mohamad Eid
Jongeun Cha

Haptics Technologies

Bringing Touch to Multimedia

 Springer

Abdulmotaleb El Saddik
Mauricio Orozco
Mohamad Eid
Jongeun Cha
University of Ottawa
School of Information Technology and Engineering
800 King Edward Ave.
Ottawa, ON K1N 6N5
Canada
elsaddik@mcrlab.uottawa.ca
morozco@discover.uottawa.ca
meid@site.uottawa.ca
jcha@discover.uottawa.ca

ISSN 2192-2977   e-ISSN 2192-2985
ISBN 978-3-642-27031-4   ISBN 978-3-642-22658-8 (eBook)
DOI 10.1007/978-3-642-22658-8
Springer Heidelberg Dordrecht London New York

Cover design: deblik

Printed on acid-free paper

Springer is part of Springer Science+Business Media (www.springer.com)

For my sweethearts for the joy they bring me.
To my fellow learners (teachers and students)
in the lifelong learning journey.
  – Abdulmotaleb El Saddik

  To my parents, Maria and my daughters
Valeria and Ximena.
  – Mauricio Orozco

  To my family and all the young people who
are struggling for freedom and peace.
  – Mohamad Eid

  To my family
  – Jongeun Cha

Preface

The first sense that a baby will experience is touch. The feelings of warmth, cold, roughness, softness, and hardness are those that we as babies first experienced and responded to.

An Australian woman gave birth to twins but the boy, thirteen weeks premature and weighing only 2lbs, was not breathing. Despite the desperate attempts of the attending medical staff to resuscitate the tiny child, after twenty minutes the doctor handed the baby to the heartbroken mother so that she could have a final cuddle and say her goodbyes.

The mother held the little boy against her skin snuggling, stroking and talking to him while her husband comforted her. Amazingly, after holding the child for two hours, she felt a slight movement from the little body. Gazing at the child she saw that he was breathing and his eyes started to open. The doctor and nurses were at first convinced that the movement was simply the muscular reaction that often occurs in a body after death. The mother, undeterred, moistened the baby's lips with a little breast milk on her finger and was overjoyed when the child tried to suck her finger. The medics were stunned but were later told that this phenomenon is not unique and is known as the 'kangaroo touch'. Being held 'skin to skin', the baby was revived by the warmth of the mother's flesh, at exactly the same temperature, and the feel of her heart beating [Sun Newspaper, August 25, 2010].

Children need affection! Imagine a child, somewhere in this world, for some reason lost connection with his/her loved ones. The traditional ways to maintain such a connection might be simple images or maybe some audio files of their voices. Would it not be exciting to be able to restore the smell, touch, and hug feeling of the child's parent whenever he/she needs their affection? Would it not be amazing to share the parent's physical affection while viewing their picture or hearing their voice recording? Would it not be interesting to recall past memories of childhood by recording and later replaying such physical stimuli?

These ideas have triggered new research into ways of physically recording those expressions of affection. This research covers methods, algorithms, and technologies for understanding, capturing, and transmitting these expressions in a realistic and secure manner.

This book is about haptics as the new media. It describes human haptic perception and interfaces and presents fundamentals in haptic rendering and modeling in virtual environments. The book explains the diverse software architectures for standalone and networked haptic systems. It also demonstrates the vast application spectrum of this emerging technology together with its trends. The primary objective is to provide a comprehensive overview and a practical view of haptic technologies. An understanding of the close relationship among the wide range of disciplines that constitute a haptic system is a key principle toward the successful building of collaborative haptic environments.

This book is different than any other book that has looked at haptics. We look at haptics as a new medium rather than just a domain in human–computer interaction, virtual reality, or robotics. It is structured as a reference book, so it allows for fast accommodation to most of the issues concerned. It is also intended for researchers interested in studying touch and force feedback for use in technological multimedia systems in computer science, electrical engineering, or other related disciplines. Many are searching for the next big haptic idea in research and development areas such as military, gaming, or interpersonal communication.

Abdulmotaleb El Saddik
Mauricio Orozco
Ottawa   *Mohamad Eid*
Jongeun Cha

Acknowledgments

Knowledge is understanding that a tomato is a fruit. *Wisdom* is not putting it in a fruit salad.

This book can also be described as a long journey since it took 5 years from the idea to the realization. It would not have been completed without the constant support, inspiration, and sacrifice of our families.

A passport of a great journey is usually stamped with funny and memorable moments. Those most beautiful memories happened either in our research laboratories at the University of Ottawa or during our many trips. Having said this, we are particularly thankful of our research fellows at the Multimedia Communications Research Laboratory (MCRLab) and the Distributed & Collaborative Virtual Environments Research Laboratory (DISCOVER). Their research and development helped us to understand a lot of the concepts presented in this book. This is due to the great environment we have in the Faculty of Engineering at the University of Ottawa. In particular, we thank Prof. Emil Petriu.

Special thanks go to Dan Martin, Hussein Alosman, and Genvieve Freeman, who sometimes under extreme time constraints were ready to proofread this work and provide us with suggestions and help. Dan, we thank you for some sleepless nights, particularly during the last few weeks of this journey. Ralf Gerstner from Springer-Verlag also deserves special thanks for his patience while working with us.

During the journey of writing this book, although we were based at the University of Ottawa, we had the opportunity to visit many places for varying lengths of time as visiting researchers or visiting professors. We greatly acknowledge the help and support we received by NSERC, Ontario Centre of Excellence, Ontario Research Fund, Canada Foundation for Innovation, Humboldt Foundations, Consejo Nacional Ciencia y Tecnología (CONACyT), TU Darmstadt, UC III de Madrid, NYUAD, as well as some private hosts during our journey in Seoul, Mexico City, and Beirut. A special acknowledgment goes to Air Canada for providing wireless access in their lounge and for the crew's hospitality while working on this book in the air.

Ralf Steinmetz and Alejandro Garcia-Alonso were the dynamos who were always asking about the status of this book and pushing for it to get finished. "Muchas Gracias"!

Finally, Nicolas Georganas: you were such an inspiring person. We owe you a lot. May your soul rest in peace.

Contents

Chapter 1
Haptics: General Principles

1.1 Introduction

Our senses are physiological tools for perceiving environmental information. Humans have at least five senses (as defined and classified by Aristotle). These senses are: sight or vision, hearing or audition, smell or olfaction, touch or taction, and taste or gustation. They are perceived when sensory neurons react to stimuli and send messages to the central nervous system. We actually have more than five senses. For example, Gibson has stated that we have both outward-orientated (exteroceptive) senses and inward-orientated (interoceptive) senses [127]. The sense of equilibrium, also known as proprioception, is one example of these other senses. Each of the sense modalities is characterized by many factors, such as the types of received and accepted data, the sensitivity to the data in terms of temporal and spatial resolutions, the information processing rate or bandwidth, and the capability of the receptors to adapt to the received data.

1.2 Human Senses

Typically, it is believed that vision and audition convey the most information about an environment while the other senses are more subtle. Because of this, their characteristics have been widely investigated over the last few decades by scientists and engineers, which has led to the development of reliable multimedia systems and environments.

1.2.1 Vision

The visual sense is based on the level of absorption of light energy by the eye and the conversion of this energy into neural messages. The acceptable wavelength range for

A. El Saddik et al., *Haptics Technologies*, Springer Series on Touch and Haptic Systems, DOI 10.1007/978-3-642-22658-8_1, © Springer-Verlag Berlin Heidelberg 2011

human eyes is between 0.3 and 0.7 μm (1 μm $= 10^{-6}$ m). The temporal resolution sensitivity of the human visual system is biologically limited and not sufficient to detect the presentation of sequential video frames past a certain speed. This is the reason why we do not perceive a digital movie as a series of still images, but rather as moving pictures. Similarly, our spatial resolution is limited and does not allow us to resolve individual pixels. The spatial resolution is determined by the density and type of photoreceptors in the retina. Several factors limit the retina's functionality, such as the size of the pupil, the stimulated area of the retina, the eye movement, the background light, and the exposure time of the target.

1.2.2 Audition

The human auditory system transmits sound waves through the outer, middle, and inner ears. This sound wave is transformed into neural energy in the inner ear. It is then transmitted to the auditory cortex for processing. The audible frequency of humans ranges from 16 to 20,000 Hz and is most efficient between 1,000 and 4,000 Hz. A sound can also be described in terms of the sound wave's direction (or relative position of the emitter to the receiver since each ear has a nonuniform directional sensitivity), frequency, intensity, or loudness (which ranges from 0 to 160 dB), and duration.

1.2.3 Touch

Indeed, the sense of touch is distributed over the entire body, unlike the other conventional four senses, which are centralized around specific parts of the body. The sense of touch is mainly associated with active tactile senses such as our hands. Such senses can be categorized in several ways, and they have a link to the kinesthetic senses. Humans are very sensitive to touch, but different parts of our body have different sensitivities. These sensitivities vary because the skin is an interface that centrally discriminates four modalities of sensation, namely touch (including both light touch and pressure), cold, heat, and pain. Furthermore, a combination of two or more modalities can be used to characterize sensations such as roughness, wetness, and vibration. A human would not be able to sense and respond to the physical environment without these tactile receptors located over the entire body. To appreciate the sense of touch more fully, consider the following facts: according to Heller and Schiff [155], touch is twenty times faster than vision, so humans are able to differentiate between two stimuli just 5 ms apart; Bolanowski et al. [44] found that touch is highly sensitive to vibration up to 1 KHz, with the peak sensitivity around 250 Hz; and skin receptors on the human palm can sense displacements as low as 0.2 μm in length [197].

1.2.4 What Does the Sense of Touch Do for Us?

Robles de la Torre [309] states that losing the sense of touch has catastrophic effects such as impairment of hand dexterity, loss of limb position perception, and the inability to walk, just to name a few. Every day we use human–computer interfaces to interact, communicate, or perform various tasks, e.g., sending e-mails, downloading a video, controlling a process in an industrial plant. It seems that audio and visual feedback is dominant for these types of interactions; however, there is considerable importance in developing and applying sophisticated touch-enabled interfaces to perform similar tasks or improve the performance of existing tasks. Therefore, the following question may arise: what level of realism can be achieved upon enabling touch interactions with virtual environments? To answer this question, the haptic modality must be more fully explored [310].

1.3 Haptics Exploration

Haptics, a term that was derived from the Greek word "haptesthai" meaning "of or relating to the sense of touch," refers to the science of manual sensing (exploration for information extraction) and manipulation (for modifying the environment) through touch. It has also been described as "the sensibility of the individual to the world adjacent to his body by the use of his body" [127]. This word was introduced at the beginning of the twentieth century by researchers in the field of experimental psychology to refer to the active touch of real objects by humans. In the late 1980s, the term was redefined to enlarge its scope to include all aspects of machine touch and human–machine touch interaction. The 'touching' of objects could be done by humans, machines, or a combination of both, and the environment can be real, virtual, or a combination of both. Also, the interaction may or may not be accompanied by other sensory modalities such as vision or audition. Currently, the term has brought together many different disciplines, including biomechanics, psychology, neurophysiology, engineering, and computer science, that use this term to refer to the study of human touch and force feedback with the external environment.

Touch is a unique human sensory modality in contrast with other modalities. As previously mentioned, it enables bidirectional flow of energy due to the sensing and acting activities performed, as well as an exchange of information between the real or virtual environment and the end user (see Fig. 1.1). This is referred to as active touch. For instance, to sense the shape of a cup, one must run his/her fingers across its shape and surfaces to build a mental image of the cup. Furthermore, in a manipulation task, for instance sewing with a needle, the division between "input" and "output" is often very sharp and difficult to define. This co-dependence between sensing and manipulating is at the heart of understanding how humans can so deftly interact with the physical world.

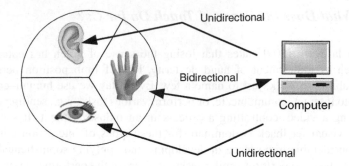

Fig. 1.1 A distinguishing feature of haptics is the bidirectional flow of information

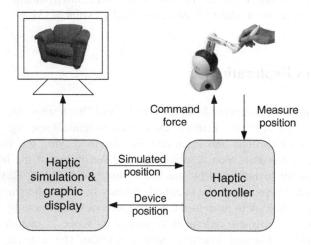

Fig. 1.2 Force representation in a virtual world

The initial sense of contact when one's hand interacts with an object is provided by the touch receptors (nerves endings) in the skin. The receptors provide information on the geometry, texture, slippage, etc. of the object surface. This information is tactile or cutaneous. When the hand applies force, trying to hold this object, kinesthetic information (force feedback) comes into play by providing physical information about the position and motion of the hand relative to the object (see Fig. 1.2).

From Fig. 1.2, one can see how we can make objects that populate the virtual environment touchable. The basic principle behind haptic interaction is simple. When the human user manipulates the generic probe (sometimes referred to as end-effector) of the haptic device, the position sensors of the device convey its tip position to the computer. At every time interval – say every 1 ms – the computer that controls the device checks for collisions between the simulated stylus and the virtual objects populating the virtual environment. If a collision has occurred, the haptic rendering system calculates the reaction forces/torques that must be applied at the human–device interaction point and controls the actuator (a computer controlled electric DC motor) attached to the device, leading to a tactual perception of the virtual objects. In the case that no collision is detected, no forces will be computed/applied, and the user is free to move the stylus as if exploring empty space. In the simplest case, the magnitudes of the reaction forces are assumed proportional to the depth of indentation, and the forces are applied immediately following surface penetration.

1.4   Concepts and Terminology

We rely on our sense of touch to do everyday tasks such as dialing a touch-tone phone, finding first gear in a manual transmission car, or playing a musical instrument. We rely heavily on the tactile and kinesthetic cues we receive from the environment. Tactile cues include textures, vibrations, and bumps, while kinesthetic cues include weight, impact, etc. In the following section, we present some crucial concepts and terminology related to haptics:

Haptic: the science of applying tactile, kinesthetic, or both sensations to human–computer interactions. It refers to the ability of sensing and/or manipulating objects in a natural or synthetic environment using a haptic interface.

Cutaneous: relating to or involving the skin. It includes sensations of pressure, temperature, and pain.

Tactile: pertaining to the cutaneous sense, but more specifically the sensation of pressure rather than temperature or pain.

Kinesthetic: relating to the feeling of motion. It is related to sensations originating in muscles, tendons, and joints.

Force Feedback: relating to the mechanical production of information that can be sensed by the human kinesthetic system.

Haptics or haptic technology: an emerging interdisciplinary field that deals with the understanding of human touch (human haptics), motor characteristics (machine haptics), and with the development of computer-controlled systems (computer haptics) that allow physical interactions with real or virtual environments through touch.

Haptic communication: the means by which humans and machines communicate via touch. It mostly concerns networking issues.

Haptic device: is a manipulator with sensors, actuators, or both. A variety of haptic devices have been developed for their own purposes. The most popular are tactile-based, pen-based, and 3 degree-of-freedom (DOF) force feedback devices.

Haptic interface: consists of a haptic device and software-based computer control mechanisms. It enables human–machine communication through the sense of touch. By using a haptic interface, someone can not only feed the information to the computer but can also receive information or feedback from the computer in the form of a physical sensation on some parts of the body.

Haptic perception: the process of perceiving the characteristics of objects through touch.

Haptic rendering: the process of calculating the sense of touch, especially force. It involves sampling the position sensors at the haptic device to obtain the user's position within the virtual environment. The position information received is used to check whether there are any collisions between the user and any objects in the virtual environment. In case a collision is detected, the haptic rendering module will compute the appropriate feedback forces that will finally be applied onto the user through the actuators (see Fig. 1.2). Haptic rendering is, therefore, a system that consists of three parts, a collision detection algorithm, a collision response algorithm, and a control algorithm.

Sensors and Actuators: a sensor is responsible for sensing the haptic information exerted by the user on a certain object and sending these force readings to the haptic rendering module. The actuator will read the haptic data sent by the haptic rendering module and transform this information into a form perceivable by human beings.

Tele-haptics: the science of transmitting haptic sensations from a remote explored object/environment, using a network such as the Internet, to a human operator. In other words, it is an extension of human touching sensation/capability beyond physical distance limits.

Tele-presence: the situation of sensing sufficient information about the remote task environment and communicating this to the human operator in a way that is sufficient for the operator to feel physically present at the remote site. The user's voice, movements, actions, etc. may be sensed, transmitted, and duplicated in the remote location. Information may be traveling in both directions between the user and the remote location.

Virtual Reality (VR): can be described as the computer simulation of a real or virtual (imaginary) world where users can interact with it in real time and change its state to increase realism. Such interactions are sometimes carried out with the help of haptic interfaces, allowing participants to exchange tactile and kinesthetic information with the virtual environment.

Virtual environment (VE): is an immersive virtual reality that is simulated by a computer and primarily involves audiovisual experiences. Despite the fact that the terminology is evolving, a virtual environment is mainly concerned with defining interactive and virtual image displays.

Collaborative virtual environments (CVE): is one of the most challenging fields in VR because the simulation is distributed among geographically dispersed computers. Potential CVE applications vary widely from medical applications to gaming.

Simulation engine: is responsible for computing the virtual environment behavior over time.

Collaborative haptic audio visual environment (C-HAVE): in addition to traditional media, such as image, audio, and video, haptics – as a new media – plays a prominent role in making virtual or real-world objects physically palpable in a CVE. A C-HAVE allows multiple users, each with his/her own haptic interface, to collaboratively and/or remotely manipulate shared objects in a virtual or real environment.

1.5   Roadmap to Multimedia Haptics

In a virtual environment, a real scenario is simulated by a computer generated application where some of the user's senses are ingeniously represented in order for them to interact and perceive stimuli that are very similar to the real environment. Traditionally, human–computer interfaces have delivered types of stimuli that are based on two of our senses, namely vision and sound. However, with the addition of the sense of touch through tactile and force feedback, the computer-based applications become richer in media content through better mimicry of real-life situations and tasks or remote real environments.

The sensing of forces is tightly coupled with both the visual system and one's spatial sense; the eyes and hands work collectively to explore and manipulate objects. Moreover, researchers have demonstrated that haptic modality reduces the perceived musculoskeletal loading that is measured through pain and discomfort in completing a task [92]. Therefore, there is a trend in the design of interfaces toward multimodal human–computer interaction that incorporates the sense of touch.

Our perceptions of the world arise as a combination of correlated inputs across several of our senses. With this in mind, we might ask whether it is possible to increase our sensory perception by simultaneously coupling visual cues to the haptic modality in a haptic-based application. In literature, it is found that most haptic-based applications, with the exception of those designed for the visually impaired, seem to be augmented by visual feedback. Many researchers have shown that the interaction with stimuli arriving in more than one sensory modality can increase the realism of a virtual reality. However, the keyword here is "perception", so if the cross-modal information is not well synchronized and consistent, the added sensory information might corrupt the intended stimulus. For instance, researchers have found that when conflict between sensory cues (for instance, between the hands and eyes) arise, the brain effectively splits the difference to produce a single mental image, and the overall perception experienced by the subject will be a compromise between the two senses. Therefore, visual cues must be synchronized with haptic interactions to increase the quality of perception.

It would be easier to extract shape information through visual means than to collect this information haptically. Exploring an object to perceive its shape using the sense of touch places large demands on the observer's memory for the exploration and integration of spatial and temporal signals. In contrast, the optimal exploratory procedures for texture – pressure and lateral motion – are simple and quick to perceive using a haptic modality. Therefore, visual cues help us anticipate the haptic sensation resulting from the interaction with an object. Imagine pressing your hand against a pillow: the visual cues have already prepared you to feel a soft object. In this case, we can say that the visual image has influenced our haptic perception.

On the other hand, many researchers have acknowledged the importance of everyday listening as: "the act of gaining information about events in the world by listening to the sounds they make" [124]. Therefore, the human auditory modality contributes intensively to the perception of the ambient environment.

In the early stages of audio-haptic inter-modal perception, it was shown that auditory stimuli do not significantly influence haptic perception [258]. Later, researchers found that sound cues that are typically associated with tapping harder surfaces were generally perceived as stiffer [95]. These studies suggest that coupling audio and haptics could help create more sophisticated perceptions of solidity, shape, location, and proximity. We believe, however, that the addition of sound to augment the perceptual capabilities of a haptic-based application is constrained by many requirements. For instance, the sound needs to be generated in real-time based on the user's interaction, and it must respond to continuous input data (such as continuous contact force). Furthermore, the synthesized sound must reflect the auditory properties of the contacting objects.

The roadmap toward Haptics Audio Visual Environments (HAVE) comprises three different paths; human haptics, machine haptics, computer haptics, and one roundabout called multimedia haptics (as shown in Fig. 1.3). Notice that the knowledge is cumulative by nature. For instance, to design a proper haptic device, one needs to understand the human haptic road, which investigates human haptic system capabilities and limitations. To develop a proper haptic rendering algorithm,

Fig. 1.3 Roadmap to multimedia haptics

one needs a knowledge of spatial and temporal attributes of haptic devices, which lies in machine haptics, etc.

1.5.1   Path 1: Human Haptics

Human haptics refers to the study of human sensing and manipulation through tactile and kinesthetic sensations. When a person touches an object, the interaction force or pressure is imposed on the skin. The associated sensory system conveys this information to the brain, which leads to perception. As a response, the brain issues motor commands to activate the muscles, which results in hand or arm movements. Human haptics focuses mainly on this human sensorimotor loop and all aspects related to human perception of the sense of touch. Therefore, human haptics research deals with all the mechanical, sensory, motor, and cognitive components of the body–brain haptic system.

Haptic perception can be defined as the process of interpreting touch information, or the sense of feeling things via the sense of touch, to recognize objects. It involves tactile perception through the skin and kinesthetic perception through the movements and positions of the joints and muscles. Humans explore and identify an object by moving their fingers on the object's surface or by holding and moving the whole object, which is called haptic perceptual exploration, and it is identified as active touch as opposed to passive touch [310].

The journey toward multimedia haptics starts by understanding the human haptic system, including the tactile and kinesthetic perceptual processes and the functioning of the human perceptual system. Researchers in this domain strive to comprehensively understand the human haptic system. This includes research into understanding the human sensory system, haptic perception and cognition in the human brain, and the human motor system (actuation system). This research also provides guidelines for the design and development of haptic interfaces. Chapter 3 of this book thoroughly covers the fundamental concepts and state-of-the-art research in human haptics.

Once the physiological elements needed to reproduce the real world as a virtual scenario have been identified, we turn to the discipline that covers such requirements in practical terms. The discipline of developing haptic technology has been named "machine haptics".

1.5.2   Path 2: Machine Haptics

Based on the knowledge of the capabilities and limitations of the human sense of touch, the second phase is to design and develop haptic interfaces – or what is referred to as machine haptics. Machine haptics involves designing, constructing, and developing mechanical devices that replace or augment human touch. These devices, also called haptic interfaces, are put into physical contact with the human body for the purpose of exchanging (measuring and displaying) information with the human nervous system. In general, haptic interfaces have two basic functions; first, they measure the poses (positions and/or orientations) and/or contact forces of any part of the human body, and second, they display the computed reaction touch to a haptic scene that populates touchable virtual objects with haptic properties such as stiffness, roughness, friction, etc. Haptic interfaces can be broadly divided into two categories: force feedback devices and tactile devices. Force feedback devices display force and/or torque and enable users to feel resistive force, friction, roughness, etc. Tactile devices present vibration, temperature, pressure, etc. on the human skin and display textures of a virtual object or provide information such as showing direction, reading text, displaying distance, etc.

Force feedback devices behave like small robots that exchange mechanical energy with users. One way to distinguish between haptic interface devices is by the number of DOFs of motion and/or force present at the device–body interface. Devices with three to six DOFs are mostly used because of their mechanical and programming simplicity in addition to their low cost. The users usually grab and move the device, which controls a tool-type avatar in a haptic scene, and when the avatar makes contact with an object in the scene, the contact force and/or torque is displayed to the user's hand through the device. Multi-DOF force feedback devices such as hand-worn gloves and arm-worn exoskeletons can provide more dexterity but are usually bulky and hard to wear. Combining multiple low-DOF force feedback devices provides simplicity and dexterity such as in two-finger grabbing. Another possible classification of force feedback devices relates to their grounding locations. Two examples are ground-based and body-based. Finally, the desirable characteristics of force feedback devices include, but are not limited to, the following: (1) symmetric inertia, friction, stiffness, and resonant-frequency properties, (2) balanced range, resolution, and bandwidth of possible sensing and force reflection, and (3) low back-drive inertia and friction [329].

Tactile devices are arrays of actuators that have direct contact with human skin. Since an actuator module cannot cover the entire continuous surface of the specific human body part, and since human skin cannot distinguish two adjacent stimuli within a certain threshold (two-point threshold) [316], most tactile devices consist of a number of actuator modules that are uniformly distributed. As discovered through human haptics research, the human body has various two-point thresholds across the body, so the density of the actuators is dependent on these thresholds. For example, the fingertip has a very small two-point threshold compared to that of the arm, so fingertip tactile devices have finely distributed actuators compared to armband-type tactile devices. Tactile devices are also broadly categorized by the stimuli that they can generate, whether it is vibration, pressure, temperature, etc., and they can be further categorized by their actuator types, such as pneumatic, motor, hydraulic, shape memory alloy, etc. Since tactile devices provide cutaneous stimuli while force feedback devices provide kinesthetic stimuli, these two types of devices can be combined to provide a very natural haptic feedback.

In a nutshell, this path involves researchers acquiring knowledge about the existing sensory and actuation hardware technologies and the control of such devices. Researchers are concerned about the design and implementation of efficient and effective sensors and/or actuators that make up a haptic device. Furthermore, this domain explores the attributes that define the quality of haptic interfaces that are based on electromechanical technologies. This domain is extensively presented in Chap. 4 of this book along with a taxonomy of haptic interfaces according to the proposed quality attributes.

Today almost any electromechanical interface requires a human–machine interface, which enables the user to interact with the simulated or remotely located real world. These devices are mainly products of research and development on computational elements related to computer haptics.

1.5.3   Path 3: Computer Haptics

Once haptic interfaces are developed, we move from the machine haptics path to the computer haptics path. Computer haptics is related to the design and development of algorithms and software that compute interaction forces and simulate physical properties of touched objects, including collision detection and force computation algorithms. Essentially, computer haptics deals with modeling and rendering virtual objects for real-time display by touch, and this computing process is called haptic rendering; it is analogous to graphic rendering. We anticipate rapid improvements in computer haptics as computers become more powerful and affordable and sophisticated software tools and techniques become increasingly available.

Since the term haptic rendering has been widely used in literature with slightly different meanings, we explicitly define it as the following:

"Haptic rendering refers to the set of algorithms and techniques that are used to compute and generate forces and torques in response to interaction between the haptic interface avatar inside the virtual environment and the virtual objects populating the environment."

The above definition has many implications. First, the avatar is a virtual representation of the haptic interface whose position and orientation are controlled by the operator. The avatar's geometry and type of contact varies according to the application and can be point based (3-DOF), object based (6-DOF), multipoint based (multiple 3-DOF), or volumetric based. The point-based haptic interface is the most widely used interface since it is computationally efficient for presenting stable haptic interaction and provides pen-like tool-based interaction that allows the user to perform a variety of tasks. Although object-based avatars give more realistic interaction forces and torques, such tools require computationally expensive algorithms; such computations, if not completed promptly, can cause the device to become unstable. These haptic devices are not easily available due to their bulkiness and high cost. An alternative to this is the multipoint-based representation. This is simply a set of multiple point-based representations that is used to provide grabbing functionality and allow more dexterous interactions such as two-finger grabbing. Volumetric-based representation is usually found in medical applications to enable cutting, drilling, etc., by sacrificing memory storage for accelerated complex and time-consuming computations.

The second implication is that the ability to find the point(s) of contact is at the core of the haptic rendering process. This is the problem of collision detection, which becomes more difficult and computationally expensive as the complexity of the models increases. However, an important characteristic of haptic interaction, locality, drastically accelerates the collision detection process by building hierarchical bounding volumes. While graphic rendering occurs globally to display the whole viewing area, haptic interaction happens in the vicinity of the haptic interface avatar, so the collision needs to be examined only around the avatar. By hierarchically dividing virtual objects into bounding volumes, the only virtual objects that are examined are the ones included in the bounding volume where the haptic interface avatar is located.

The third implication is the need for a force response algorithm. This is the calculation of the ideal contact forces. Upon detecting a collision in a virtual environment, interaction forces between avatars and virtual objects are computed and transmitted to users via haptic interfaces, generating tactile and/or kinesthetic sensations. The interaction force is generally calculated based on a penetration depth, described as the distance the haptic interface avatar penetrates the object it is acting upon. Due to the mechanical compliance of haptic interfaces and the discrete computation characteristics of computers, the haptic interface avatar often penetrates virtual objects [410]. By introducing an ideal haptic interface avatar that has the same position as the haptic interface avatar in free space and cannot penetrate virtual objects, namely a god-object or a proxy, the penetration depth is calculated as the distance between the real haptic interface and ideal haptic interface avatar. As a result, the interaction force is calculated according to Hooke's law[1] using the stiffness value of the virtual object being acted upon. In order to add surface properties such as friction or roughness to the calculated force, the position of the ideal haptic interface avatar on the virtual object can be modulated.

The final implication is that the interaction between avatars and virtual objects is bidirectional; the energy and information flows both from and toward the user. This means that the virtually generated energy in the virtual environment is physically embodied via haptic interfaces and can potentially injure the user or pose a safety problem. Generally, this can be avoided by keeping the haptic rendering update rate higher than 1 kHz, providing a reasonable amount of stable and smooth force to simulate stiff objects [53]. However, in order to guarantee stability of haptic rendering in low power systems, or to keep high fidelity, haptic control algorithms need to be considered as introduced in [81, 147].

---

[1]$F = kx$, where F is the restoring force, x is the penetration depth, and k is a stiffness value of the closest surface.

Consequently, computer haptics provides software architectures for haptic inter-actions and synchronization with other display modalities. Chapter 5 of this book presents the fundamental concepts of haptic rendering along with some discussions about design and implementation details.

1.5.4 The Roundabout: Multimedia Haptics

The last phase in this journey is multimedia haptics, which considers haptics as a new media channel in a complete multimedia system. We define multimedia haptics as the following:

"the acquisition of spatial, temporal, and physical knowledge of the environment through the human touch sensory system and the integration/coordination of this knowledge with other sensory displays (such as audio, video, and text) in a multimedia system"

Multimedia haptics involves integrating and coordinating the presentation of haptic and other types of media in the multimedia application. Generally, a multimedia system consists of media acquisition or creation, content authoring, and transmission and consumption components. Multimedia haptics research can be categorized based on these components as described below [69].

First of all, haptic content needs to be created before it can be delivered to the audience and consumed. While there are a lot of standard tools to capture or synthesize audio and video (AV) media, such as a camcorder, it is less obvious how the same objective can be achieved for haptic media. Basically, haptic media can be created through three key approaches like what has been done in AV media: data can be recorded using physical sensors; it can be generated using specialized modeling tools; and it can be derived automatically from analysis of other associated media.

The acquired haptic media needs to be represented in a proper format so that it can be stored synchronously with other media. There have been endeavors to add haptic media into existing multimedia representation frameworks such as Reachin API to VRML, H3D into X3D, HAML based on XML, and haptic broadcasting framework based on MPEG-4. Furthermore, MPEG-V Media Context and Control (ISO/IEC 23005) is another framework that deals with sensory information, including haptic/tactile modality in a virtual world. Haptic media can be synchronized temporally and spatially with the multimedia representation to produce meaningful content. This requires haptic authoring tools, which are counterparts of audiovisual media production tools such as video authoring tools, 3D modeling tools, etc. 3D haptic modelers, such as HAMLAT (HAML-based Authoring Tool [103]) and K-Touch [336], provide graphic and haptic user inter-faces for adding haptic properties to existing virtual objects. Tactile editors, such as posVibEditor [322], a tactile movie authoring tool [206], enable the creation and editing of vibration patterns that can be synchronized with AV media. We call the resultant authored content "haptic content" to differentiate it from AV content.

The generated haptic content can be stored in files to be delivered through storage devices such as CD/DVD, USB memory drive, etc. or transmitted through the communication network. Sometimes the haptic content may be acquired and transmitted immediately for real-time interactions in a shared virtual simulation. Traditionally, the implementation of the shared virtual simulation is limited by two problems: latency and coherency in manipulation. Delays in processing haptic media can easily bring the haptic interface to a state of instability. Therefore, intense research has been undertaken to reduce delays and jitters in processing and transmitting force information over long distances. The various techniques that were developed to integrate force feedback in shared virtual simulations must deal with significant and unpredictable delays and synchronization issues. One example of such a system is the Collaborative Haptic Audio Visual Environments (C-HAVE), which allows multiple users with their own haptic interfaces to collaboratively and/or remotely manipulate shared objects in a virtual environment.

To recapitulate, this phase covers concepts such as haptic data capturing and representation, transmission and compression, and the synchronized dissemination of haptic media have been explored. One of the most challenging areas of research in haptics is the on-time communication of haptic data, and currently, extensive research is being conducted in the domain of haptic media transmission (or tele-haptics). Several communication frameworks and protocols for haptic data communication, as well as performance issues and challenges, will be discussed in Chap. 6 of this book.

1.6 Haptic-Audio-Visual Multimedia System

With the technologies that have been developed in human haptics, machine haptics, computer haptics, and the multimedia haptics, conventional multimedia systems have the potential to evolve into a haptic-audio-visual (HAV) multimedia system that brings more interactive and immersive experiences because of the haptic modality. For example, while viewers passively watch TV or movies, or gamers interact with video game characters audiovisually through a game controller, the HAV multimedia users would be able to touch the game characters and feel a physical event, such as an earthquake, happening in a movie. Figure 1.4 shows a diagram of the HAV multimedia system; compared to the conventional multimedia system, haptic sensors and displays are added to capture haptic properties and display haptic interactions through corresponding devices. Since most haptic displays have physical contact with the human body, they are designed based on human haptics and machine haptics to provide a comfortable and safe interaction experience. The interaction response, such as a reacting force or vibration on the skin, is simulated and calculated via the haptic rendering process, which works together closely with the graphic rendering process and other simulation engines. In order to provide a stable and high fidelity interaction force, high update rates need to be maintained,

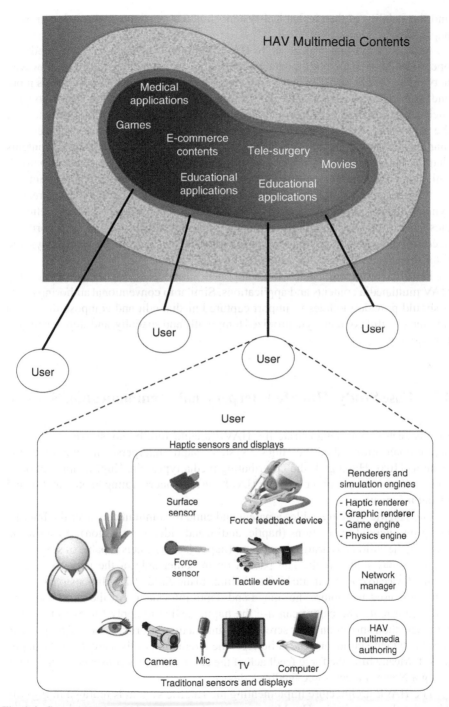

Fig. 1.4 General HAV multimedia system

and the stability of the mechanical system should be guaranteed through computer haptics.

Another important aspect in a HAV multimedia system is their mode of operation. Most HAV multimedia contents work in a standalone fashion; however, networked collaborative environments have gained interest in a society that is more and more interconnected these days. Thus, a haptic transmission over networks such as a dedicated network or the Internet can be located at different places in the architecture based on computer haptic and multimedia haptics. The network management component is responsible for communicating the multimedia contents (haptic, audio, visual, etc.) over a network (dedicated or nondedicated networks). This component implements all network related algorithms to maintain intramodal (within the same media stream) and intermodal (between different media streams) synchronization for the multimedia contents and compensates for network deficiencies such as network reliability and delay/jitter. Several specific haptic algorithms need to be developed to compensate for information loss and network delays and jitter to maintain the stability of the haptic interactions.

The HAV multimedia authoring component allows users or producers to develop HAV multimedia contents and applications. Similar to conventional authoring tools, it should provide modules to import captured media, edit and compose the media into meaningful contents synchronized temporally and spatially, and store the results for delivery.

1.7   Case Study: HugMe Interpersonal Communication System

In this section, we demonstrate the HAVE architecture by considering a specific haptic application called the HugMe system (haptic interpersonal communication system) [105]. Having all the contributing media types, the HugMe application is an excellent example of a complete HAVE system incorporating haptic, audio, and visual information.

The HugMe system enables a parent and child to communicate over the Internet using multimodal interactions (haptic, audio, and video information). As shown in Fig. 1.5, the child is wearing a haptic suit (haptic jacket) that is capable of simulating nurturing physical stimuli. The parent, on the other side of the network, uses a haptic device to communicate his/her feelings to the child. A 2.5-dimensional (2.5D) camera is used to capture the image and depth information of the child and send it to the parent. The parent can use the haptic device to apply forces to the child representation shown on their screen. The interaction information is calculated and sent back to the child, and the child feels the interaction of the parent via the haptic jacket. Meanwhile, the force feedback of the child's image is conveyed to the parent using a Novint Falcon force feedback device.

The HAVE architecture implementing the HugMe system is redrawn in Fig. 1.6. The visual sensor in the HugMe system is the depth video camera that captures the color information (RGB signal) and depth information (D signal). The depth

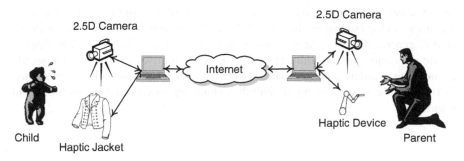

Fig. 1.5 HugMe application scenario

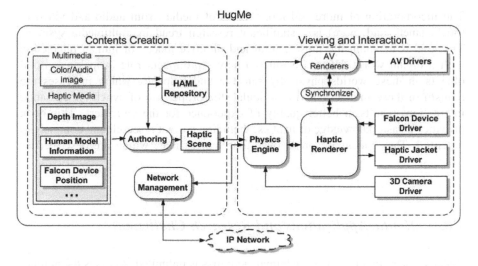

Fig. 1.6 HugMe system architecture

signal is a grayscale bitmap image where each pixel value represents the distance between the camera and the respective pixel in the RGB image. The HugMe system uses a commercially available camera, called the ZCam^{TM} from 3DV Systems, to capture both the RGB and the depth data. Furthermore, special markers are used to track the child's body movements and construct a human model that is used in collision detection. All the captured information is stored in a data repository using the HAML format. The network management component implements Admux (a multimedia communication protocol for synchronous haptic-audio-video communication) [104], which synchronizes the multimedia rendering and adapts it to the network requirements, compensating for any network deficiencies.

The haptic interface used in the HugMe system is the Falcon device, developed and marketed by Novint Technologies, Inc. It provides the parent with the touch feeling whenever the haptic device end-effector collides with an object of the remote environment (in our case the video contents). The Falcon device is both a haptic sensor (represented by the Falcon device position component) and a force feedback device (shown as the Falcon device driver component), as shown in the HAVE general diagram. At the child's side, the haptic jacket is used to display tactile information to the child. The haptic jacket comprises an array of vibrotactile actuators to simulate continuous tactile feeling to the user. The jacket is connected to the HugMe system using Bluetooth technology to enhance its mobility and wearability.

1.8   Roadmap of the Book

The incorporation of more and more forms of media (from audio and video to touch, smell, and more) is a significant research trend in multimedia systems. The goal is to attain the most natural and intuitive modes of human interaction with a digital world. Haptics is sure to play a prominent role in making virtual objects in these worlds physically sensible and palpable, which increases the realism of these interactions. In particular, the proper use of synchronous haptic interactions results in better quality of experience for the end users. This book provides a basic knowledge of haptics, including current research and commercial potential. The content is spread over four major areas, which are described as follows:

1.8.1   Haptic Applications and Research Challenges

Because the number of possible human activities is unlimited, so too is the number of haptic applications. In Chap. 2, we present a description of application categories showing some of these applications. As a matter of fact, applications of this technology have rapidly been extended to devices used in graphical user interfaces (GUIs), games, multimedia publishing, scientific discovery and visualization, arts and model creation, editing sound and images, the vehicle industry, engineering, manufacturing, tele-robotics and tele-operation, education and training, and medical simulation and rehabilitation. From the literature, one can make several observations as well as recommendations for future research in haptics for multimedia. The literature also helps to pinpoint different research challenges that the haptic community is facing; Chap. 7 does exactly this. These challenges are classified in parallel with the topics covered through the book; many stem from the limitations of haptic device hardware – impractical, expensive, and inaccessible – and the complexity of touch and physical interactions.

1.8.2 General Principles in Haptic Multimedia Systems and Human Haptic Perception

Chapters 1 and 3 explain the basic principles of a haptic system and the foundation and disciplines related to haptics. Chapter 1 introduces the basic concepts and terminology used among the haptic community and provides the "big picture" of what a haptic system is. It also explains the common features of haptic applications; the architecture of a virtual reality application that incorporates visual, auditory, and haptic feedback; and haptic perception and modeling in the virtual environment. Chapter 3 describes the biology of touch in the human body, the classification and measurement methodologies for haptic perception, and perception experimentations and tools.

1.8.3 Haptic Interfaces and Rendering

One of the most important aspects of haptic applications is the haptic interface because it provides a path for perceived stimuli and the human kinesthetic and/or touch channels. This is discussed in Chap. 4. Discussion on haptic rendering is found in Chap. 5 and covers three main topics: collision detection algorithms and their classifications, force response algorithms, and control algorithms. The collision-detection algorithm uses position information collected through sensors to find collisions between objects and avatars and report the resulting degree of penetration or indentation. Next, the force-response algorithm computes the "ideal" interaction forces between avatars and virtual objects involved in a collision. And finally, the control algorithm collects interaction force information from the force-response and applies them on the operator through the haptic device while maintaining a stable overall behavior.

1.8.4 Haptic Audio Visual Environment

In Chap. 6, we study the types and designs of applications that gain access to the virtual object perceptual information through haptic displays. Chapter 6 contains descriptions of the various techniques used in integrating force feedback into shared virtual simulations. This integration requires dealing with significant and unpredictable delays, haptic information representation, synchronization, haptic APIs, existing haptic software frameworks, such as Reachin and Novint e-Touch, and haptic programming toolkits. The chapter elaborates on the discussion of networked haptics (commonly referred to as the Collaborative Haptic Audio Visual Environment (C-HAVE)). Some characteristics, such as quality of experience and security, are also highlighted.

1.9   Further Reading about Haptics

In recent years, there has been extensive research literature on all aspects of haptic systems. Some journals, such as IEEE Transactions on Haptics, IEEE Transactions on Instrumentation and Measurement, ACM Transactions on Graphics, ACM Transactions on Applied Perception, ACM Transactions on Multimedia Computing, Communications and Applications, MIT Press Presence, Springer Multimedia Tools and Applications, Springer Multimedia Systems Journal, and the Electronic Journal of Haptics Research "Haptics-e" frequently publish haptics-based research results. Many other journals, such as Journal of Robotics and Mechatronics, have special issues on this subject.

In addition, a good number of international conferences and workshops are either dedicated to haptics or have special sessions on haptics. Some examples are: the IEEE International Symposium on Haptic Audio Visual Environments and Games (HAVE), IEEE Haptics Symposium, Eurohaptics, Worldhaptics, ACM Multimedia (ACM MM) and Human Computer Interaction (ACM HCI) Conferences, and the IEEE International Conference on Robotics and Automation.

Chapter 2
Haptics: Haptics Applications

2.1 Introduction

Mankind has always dreamt of escaping reality, and this will only be possible through the use of new technologies. According to Greek legend, when Daedalus and his young son Icarus were trapped within the Labyrinth that Daedalus had built, Icarus did not escape by finding an exit through trial and error or going through the walls; he escaped by taking to the sky using a set of wings constructed by his father. He escaped through the use of new technology. Overwhelmed with the excitement of the feeling of flying, Icarus went too high. By flying too close to the sun, the heat melted the wax that held his wings together, and he fell into the sea. Indeed, technology can be pushed too far! Escaping reality using technology, while keeping in touch with reality, is the fantasy some visionaries are trying to achieve. Some of these visionaries work on haptic audiovisual environments (HAVEs).

2.2 Haptic Evolution: From Psychophysics to Multimedia

Haptics was introduced at the beginning of the twentieth century through research in the field of experimental psychology aimed at understanding human touch perception and manipulation. These psychophysical experiments provided the contextual clues involved in haptic perception between humans and machines. The results in the disciplines of psychology and physiology provided a renewed surge into the study of haptics, and it remained popular until the late 1980s. Researchers have found that the mechanism by which we feel and perceive the tactual qualities of our environments are considerably more complex in structure than, for example, our visual modality. However, they opened up a wealth of opportunities in academic research to achieve realistic touch simulation.

Turning to the robotics arena in the 1970s and 1980s of the last century, most researchers were considering the systems aspect of controlling remote robotic

A. El Saddik et al., *Haptics Technologies*, Springer Series on Touch and Haptic Systems, DOI 10.1007/978-3-642-22658-8_2, © Springer-Verlag Berlin Heidelberg 2011

vehicles to perceive and manipulate their environments by touch. The main objective was to create devices with a dexterity inspired by human abilities. When these robotic mechanical systems have a human being in their control loop, they are referred to as tele-manipulators. In these systems, an operator is expected to perceive the environment, reason about the perceived information, make decisions based on this perception, and act according to a plan specified at a very high level [19]. In time, the robotics community found interest in topics including, but not limited to: sensory design and processing, grasp control and manipulation, object modeling, and haptic information encoding. Meanwhile, terms such as "tele-operation", "tele-presence", and "tele-robotics" were used interchangeably by the robotics community until the mid-1990s. Two of these terms ended up being especially important for haptic systems: tele-operation and tele-presence. Tele-operation refers to the extension of a person's sensing and manipulation capabilities to a remote location. In tele-presence, an operator feels as if he/she is physically at the remote site. Motivated by these concepts, the tele-presence and tele-operation research communities developed several projects in a variety of fields such as nuclear, subsea, space, and military applications. Only recently have haptic technologies become integrated with high-end workstations for computer-aided design (CAD), and at the lower end on home PCs and consoles to augment human–computer interaction (HCI). Effectively, this implies the opening of a new mechanical channel between humans and computers such that data can be accessed and literally manipulated through haptic interfaces. Currently, computer haptic systems can display objects of sophisticated complexity and behavior. This is thanks to: the availability of high-performance force-controllable haptic interfaces; affordable computational geometric modeling, collision detection, and response techniques; a good understanding of the human perceptual needs; and a dramatic increase in processing speed and memory size. With the commercial availability of haptic devices, software toolkits, and haptics-enabled applications, we foresee that the field of human–haptics interaction will experience an exciting growth. Kinesthetic movement and haptic tactile sensations allow for multimedia applications to utilize touch and force feedback in addition to traditional media such as image, audio, and video. Haptics, as a new media, plays a prominent role in making real-world objects physically palpable in a collaborative virtual environment. For instance, Collaborative Hapto Audio Visual Environments (C-HAVEs) allow multiple users, each with his/her own haptic device, to manipulate shared objects in a virtual environment.

The potential of haptics as a new media is quite significant for many applications, such as: tele-contact (haptic conference), gaming, tele-presence, tele-learning, tele-medicine, tele-operation in hazardous environments, industrial design and testing, scientific discovery and visualization, arts and creation, the automotive industry, engineering, manufacturing, education and training, as well as medical simulation and rehabilitation and any related interactive virtual reality application (as illustrated in Fig. 2.1). Therefore, the application spectrum is quite vast, and its trend of expansion is anticipated to increase. In this chapter, we provide an overview of some of the current applications involving the use of haptics as a promising technology.

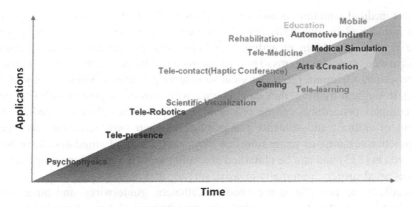

Fig. 2.1 The spectrum and trend for HAVE applications

2.3 Haptics for Medical Applications

Medicine is an ancient discipline, yet the medical field has been an active source of haptic development in the recent past. Haptics has been used in medical training to revolutionize many surgical procedures over the last few decades. Surgeons rely on the feeling of net forces resulting from tool–tissue interactions, and they require proper training to successfully operate on patients. Since haptic technology can be applied to different techniques in the medical field, we have adopted the formal medical procedure taxonomy to identify current medical-haptic-based applications. A medical procedure is "an activity directed at or performed on an individual with the object of improving health, treating disease or injury, or making a diagnosis".[1] This can include a wide variety of techniques; however, according to the Canadian Centre for Health Information [366], medical procedures can be broadly classified into diagnostic, therapeutic, and surgical procedures. While haptics can be used for education and training in diagnostic procedures, in this book we concentrate on haptics applied in surgical simulations and therapeutic procedures, particularly stroke-based rehabilitation and support for the visually impaired.

2.3.1 Surgical Simulations

Surgery can be defined as "the branch of medicine concerned with treatment of bodily injuries or disorders by incision or manipulation, especially with instruments".[2] Surgery simulation environments that utilize force feedback devices can

---

[1] International Dictionary of Medicine and Biology ISBN 047101849x.
[2] Concise Oxford Dictionary-10th Edition.

support medical simulations and training. Despite the fact that this technology has been introduced in minimally invasive surgery (MIS) procedures, among others, there is still a problem providing significant haptic feedback to improve user performance on such systems. This was pointed out by Xin et al. [398] and lately referenced by Okamura [269]. Surgical simulators potentially address many of the issues in surgical training: they can generate scenarios of varying complexity; new and complex procedures can be practiced on a simulator before proceeding to a patient or animal; and students can practice on their own schedule and repeat the practice sessions as many times as they want. Surgical simulators have been surveyed in [233] and can be classified according to their simulation complexity as needle-based surgery, minimally invasive surgery, and open surgery.

Needle-based procedures use needles, catheters, guide-wires, and other small bore instruments for teaching relatively straightforward procedures with well-defined algorithms. They are performed most commonly in abdominal surgery. The needle insertion action is sometimes difficult to perform and requires a pre-programming. A novel interactive haptic approach is presented in [96] to simulate this procedure. The virtual needle insertions are simulated using a numerical material model and a derived needle shaft force distribution. A virtual needle is advanced into a linear elastostatic model in two dimensions that are discretized using the finite element method. Other needle-based simulators can be found in [232, 383]. A similar type of simulation was established by Chial et al. [76] who presented a haptic scissor system intended to simulate the interface of a pair of Metzenbaum[3] surgical scissors. It has been tested and compared against real tissue simulations. The haptic results recorded from this project provide good guidelines for a detailed analysis for reality-based modeling, but there is still further research to be done to overcome the limitations in the presented approach.

Minimally invasive surgery uses specially designed instruments that are introduced into the body via small incisions and is commonly referred to as laparoscopic surgery. It is characterized by a limited range of motion and haptic feedback, the use of specialized tools, and video displays. Many laparoscopic simulators have been developed so far [39, 231]. For instance, a training set to simulate laparoscopic procedures based on virtual surgical instruments for deforming and cutting 3D anatomical models has been developed at the Institut National de Recherche en Informatique et en Automatique (INRIA) [288]. Their approach is based on biomechanical models that include the notion of anistropic[4] deformation.

Another framework that includes many important aspects of haptics is the Minimally Invasive Surgical Simulation and Training (MISST) framework described in [26]. Several challenges have been uncovered in the design of MIS simulators,

---

[3] American surgeon (1876–1944). The surgical scissors have been named after Dr. Myron Firth Metzenbaum. They have curved blades with blunt ends.

[4] Anistropic. Physics: having a different magnitude or properties when measured in different directions. Properties of a material depend on the direction; for example, wood. In a piece of wood, you can see lines going in one direction; this direction is referred to as "with the grain".

including the haptic interface hardware design, tissue and organ model development, tool–tissue interactions, real-time graphical and haptic rendering, and recording and playback. In the field of commercial products, Simbionix^{TM} has developed a LAP Mentor^{TM}, which is a multidisciplinary surgery simulator that offers training to both new and experienced surgeons. Training ranges from basic laparoscopic skills to performing complete laparoscopic surgical procedures [205]. LAP Mentor^{TM} provides tactile sensations transmitted by the use of laparoscopic instruments.

Open surgery requires direct vision of, and tactile contact with, a region of interest inside the human body. The visual field, the range of haptic feedback, and the freedom of motion are considerably larger compared to MIS, thus it is more difficult to simulate. For example, a biopsy is a medical procedure that relies on the manual skills of the medical surgeon. Acquisition of these skills requires gaining significant experience. One method has been proposed by Marti et al. [244] to acquire such experience through a technique that combines visualization with haptic rendering to provide real-time assistance to medical gestures. This biopsy navigator is a system that provides haptic feedback to the surgeon using patient-specific data. Realistic open surgery simulation requires considerable advances in haptics and visual rendering, real-time deformation, and organ and tissue modeling.

Simulation environments that utilize force feedback technologies can support medical education and training. The application of a desktop haptic interface for pre-operative planning and training for hip arthroplasty surgeons is introduced in [373]. Another haptic-based medical training system is introduced in [9]. Their system's Graphical User Interface (GUI) allows a trainee or an operator to simultaneously see multiple views of a virtual patient's anatomy from different perspectives. In addition, medical training for the skill of bone drilling is investigated in [112]. It was observed that enabling haptic and acoustic feedback increased the performance of the trainees and accelerated the training process. Surgeons from Pennsylvania State University's School of Medicine and Cambridge-based Boston Dynamics developed a training simulation using two PHANToM (Personal Haptic Interface Mechanism) devices [246]. Medical residents, through a simulated environment, rehearsed needle-based procedures; meanwhile, data regarding their surgical skills was collected. Many other medical education and training systems have been proposed, including a computer-based system for training laparoscopic procedures [27] and a Munich Knee Joint Simulator [307], among others.

Robot-assisted surgery has been achieved in various fields of MIS, such as appendectomies and cardiotomies (heart surgery), by the state-of-the-art da Vinci Surgical System, which became commercially available in the latter half of the 1990s and is now the most commonly used robotic system for MIS. The da Vinci system has enabled the replication of a surgeon's delicate and dexterous hand motions within the patient's body through small surgical incisions. Although the effectiveness of haptic feedback in robot-assisted tele-operated surgery has not yet been fully investigated [253], it is still evident that surgeons can benefit from haptic feedback in robotic surgery [204].

The following examples show the progress that is being made in incorporating haptics into the realm of surgery:

The authors in [38, 302] investigated the effect of visuohaptic feedback with modified da Vinci surgical instruments. Their study describes evidence that visual sensory substitution permits the surgeon to apply more consistent and precise robot-assisted knot tying and allows greater tensions in fine suture materials without breakage during the procedure.

Tholey et al. [93, 369] have developed a prototype for a laparoscopic grasper with force feedback capability, along with an information-enhanced display for providing vision and force feedback to the user while manipulating tissues. Their results confirm that with simultaneous vision and force feedback, subjects are more comfortable and more accurate at characterizing tissues compared with either vision or force feedback alone.

A tele-manipulated experimental surgical platform was developed in [93]. The platform included commercially available equivalent surgical instruments to present comparable conditions for the surgeons. One of the findings was that force feedback influences the application of forces significantly in surgical knot tying, and visual fatigue decreases significantly while operating with haptic feedback for young and conventionally experienced surgeons.

2.3.2   Stroke-Based Rehabilitation

Strokes are considered one of the leading causes of death in the world according to a World Health Organization report published in 2008. Survivors may suffer minor or major disabilities in their cognitive and motor capabilities and, as a result, are unable to carry out their usual daily activities. Typically, they enter a rehabilitation program to recover their motor abilities up to a certain extent. Virtual reality (VR) technologies have been used to provide entertaining environments for stroke patients to use as a therapeutic tool to regain fundamental motor functions. Incorporating haptic technologies into virtual environments allows patients to feel and touch the virtual environment as well.

The rehabilitation process involves applying certain forces to the injured/disabled organ (such as the finger, arm, ankle) to help it recover its strength and range of motion. Emphasis is placed on the optimization of function through the combined use of medications, physical modalities, physical training with therapeutic exercise, movement and activities modification, adaptive equipment and assistive devices, orthotics (braces), prosthetics, and experimental training approaches. Some of the mentioned techniques can benefit from the trend of current haptic technology, which, if combined with virtual environments, presents an option for optimizing current procedures. Adding force feedback information within a rehabilitation virtual environment helps to measure performance and to tailor performance-based exercises for each patient. This potential to assess a patient's performance by measuring different parameters, which cannot be evaluated in traditional rehabilitation, can be of benefit to both patients and occupational therapists.

Some relevant work that combines virtual reality and haptic technologies to handle poststroke rehabilitation of the upper and lower extremities (hand, arm, and ankle) are [42, 52, 189, 250, 280, 338, 339]. Their work demonstrates that stroke and musculoskeletal pain syndromes are the most recurrent conditions that have taken advantage of haptic technology. For example, Broeren at al. [52] proposed the use of a 3D-computer game as a training utility to promote motor relearning in a patient suffering from left arm paresis (muscular weakness). The effectiveness of haptic-guided errorless learning has also been tested in [84] with a group of twelve poststroke patients. It has been shown that the concept of using errorless learning with haptic feedback benefitted some patients, but not all. Mali and Munih [241] developed a low cost haptic device with two active degrees of freedom (DOFs) and a tendon-driven transmission system optimized for finger exercises. The device was constructed to envelop a finger workspace and to generate forces up to 10 N.

In the case of lower extremities, such as the ankle, a rehabilitation environment has been developed in [42] using the 'Rutgers Ankle' haptic device as a foot joystick. Variations in the exercises were achieved by changing the impedance levels, stiffness levels, and vibrations. The same Rutgers Ankle interface has been used in the rehabilitation of musculoskeletal injuries [94]. In keeping with the extremity areas of the body, there is a PC-based orthopedic rehabilitation system [293] and an upper limb rehabilitation system [222]. A distributed collaborative environment has also been developed in [251], which includes haptic sensory feedback, augmented with a voice conferencing system, to serve stroke patients in the subacute phase. Other applications include the rehabilitation of patients with hemispatial neglect[5] [22], hand rehabilitation, and robotic therapy using Hidden Markov model (HMM)-based skill teaching [404].

Furthermore, a haptic-based system for hand rehabilitation, consisting of a series of game-like tasks to address certain parameters of hand movement, such as grasping angles, velocities, and/or forces, was used as test-bed for experimental evaluation in [339]. The system can be set up in the patient's house to provide a treatment that is not restricted by time and/or facilities, while offering continuous evaluation of the patient's improvement. The proposed framework incorporates tests that occupational therapists have been using for a long time, such as the Jebsen Test for Hand Function (JTHF) [255] and the Box and Block test (BBT) [247].

Alamri et al. [13] presented a system based on augmented reality (AR) technology that can increase a patient's involvement in the rehabilitation exercise, and at the same time, measure the patient's performance without the direct supervision of a therapist. Their proposed system is called SIERRA, for post-Stroke Interactive and Entertaining Rehabilitation with ReActive objects. The system uses AR technology to provide a natural exercise environment containing entertaining virtual objects. It adopts a game concept to provide patients with

---

[5]"The syndrome of hemispatial neglect is characterized by reduced awareness of stimuli on one side of space, even though there may be no sensory loss" [282].

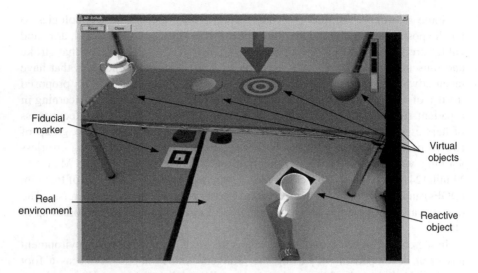

Fig. 2.2 An example of the Shelf game designed for real and virtual objects

a more entertaining environment for treatments. They seamlessly superimpose virtual objects into a real environment, which allows patients to interact with them in motivating game scenarios using a tangible object. In this system, the tangible object is three-dimensionally tracked using vision analysis algorithms, and the movement of the tangible object is mapped on a virtual avatar that can interact with the virtual environment. Since vibrotactile actuators are attached to the tangible object, the patient can experience haptic/tactile feedback in addition to audiovisual feedback. The usability study they conducted with stroke patients at the German Rehab centre in Wisbaden shows that games with vibrotactile feedback offer advantages both in terms of improving the interest of patients in the therapy and in dealing with the realism of the games, which improved the rehabilitation process. Figure 2.2 shows an example of the Shelf game, which was designed as an AR game where users make use of both real and virtual objects.

2.3.3 Support of the Visually Impaired

According to the WHO, in 2002 there were more than 161 million visually impaired people worldwide. From that figure, 124 million people had low vision while 37 million were blind (WHO 2010). In addition, age-related macular degeneration (AMD) is one of the leading causes of severe visual impairment in the aging population. Research is showing that haptics can play a vital role in improving the quality of life of those affected. A study conducted by Jacko and his team has demonstrated the benefits of multimodal components for enhancing human interactions with information technologies that have graphical user interfaces (GUIs). Their study is based on a simple drag-and-drop task using a computer mouse. Their results revealed that the inclusion of multimodal feedback including haptics improves the performance of users with AMD of varying levels of visual acuity, as well as a group of age-matched controls [190].

Haptics enhances the perception of blind or visually impaired people in applications such as learning, typing, and reading, by converting visual or sound information into a haptic modality. In the last decade, significant research has been carried out with regards to building applications and systems dedicated for blind and visually impaired people. Among the early works in this domain, Jansson realized that exploring computer generated objects for blind people has specific challenges, especially when exploring a 3D object's attributes depicted in a 2D space [192]. His study distinguished the contribution of haptic interfaces, more specifically the PHAToM device, in exploring a virtual object's physical attributes, such as roughness, by blindfolded users. His experiments demonstrated that, for those whose vision was not available, adding force feedback contributes important information to the task of exploring objects.

In his work, Levesque [224] presented a comprehensive survey of the use of haptics with the blind. A dual-point haptic interface within the European Union GRAB project (http://www.grab-eu.com) has been developed for testing three scenarios: the exploration of chart data, a city map explorer, and a simple adventure game [21]. These applications were developed and demonstrated in subsequent work by Iglesias et al. [180], where the applications were tested by visually impaired persons with different profiles, e.g. congenitally blind vs. acquired blindness, to confirm the validity and potential of the developed system.

A TACTICS system, which stands for TACTile Image Creation System, converts visual information into tactile information and was proposed in [390]. The idea is to develop an interface for visually impaired people to interpret complex scientific data. The system comprises a software/hardware architecture where graphical information is segmented and later transformed to audible components. The objective is to allow blind users to surf the web, browse a CD-ROM collection of images, or navigate a GUI with a certain degree of comprehension for the experimental set up [390]. Four experiments: simple and timed discrimination, as well as identification and comprehension of tactile images were conducted to evaluate the performance of the proposed system. For example, in an unprocessed set of tactile images 50% of the subjects performed well at the discrimination task. In addition, they found that blind subjects were 10% less accurate than sighted subjects to discriminate tactile images.

Yu et al. [405] proposed a system to support visually impaired people in accessing graphs and diagrams by exploring the outline of given objects through the sense of touch. The experimental setup was based on two different lines displayed on a graph with different friction properties. Preliminary results showed some issues in correctly identifying the layout of the line segments in question. The authors argued that the discrepancies found in the results were due to the haptic device used in the experiment (PHANToM). Furthermore, they stated that such a device is adequate for kinesthetic rather than cutaneous sensation.

"Audio Haptics for Visually Impaired Training and Education at a Distance" (AHViTED is an approach for retrieving visual information through the use of diagrams with integrated sound files (AHViTED 2010). The main goal is to improve accessibility to visual graphics by nonvisual means and to allow autonomy of use by the individual in a distance learning environment. The proof-of-concept prototype is based on different tactile technologies including touch screens on which tactile overlays are placed. A tactile overlay is laid on the touch screen, and the system is activated by the user touching the screen. In addition, when symbols, icons, and regions of the tactile surface are pressed, the user receives instant audio feedback from the computer.

The Chromo-Haptic Sensor-Tactor (CHST) is a device that represents the feeling of color. CHST contains a glove, with short-range optical color sensors mounted on its fingertips, and a torso-worn belt, on which vibrotactile actuators ("tactors") are mounted. The main purpose of developing such a device is to attribute a feeling to different colors for visually impaired people, even if they have been unsighted since birth. Recognizing colors can be critical in some scenarios (e.g., orange or red used as a hazard warning). This system was designed to map four color sensors to four vibrotactile actuators. It consists of a glove with four color-sensing modules (Red, Green, Blue and Clear) and a soft elastic fabric belt with a set of tactors. The glove and belt are interfaced to a microprocessor system. The belt can be worn around the torso in such a way that the tactors are placed against the skin with moderate pressure. The first step is to convert the sensor R, G, B, and C values to an approximate point in RGB space. Then the RGB vector is translated to a more intuitive color space. The system maps the resultant color to a vibrotactile representation by defining and transmitting tactor signals [365].

The Dynamic Tactile Map is a project that includes an Intelligent Glasses System providing stereo images and a VIbroTActiLe (VITAL) interface encompassing 8×8 vibrating microcoils that can work up to 400 Hz. The microcoils are separated from each other by a distance of 5 mm. This project presents a dynamic tactile map for dual space binary representation useful as a navigation tool for the visually impaired. The system works in steps. In the first step, the Intelligent Glasses System takes a picture of the surroundings where the user is located. Next, the image is transferred to the VITAL interface via wires. Finally, the VITAL interface represents the map of the environment as obstacles and empty space. Hence, the main purpose of implementing this system is to be used as an obstacle avoidance system [240]. The Tactile Handle is a device used to direct blind people through both familiar and unfamiliar environments without relying on the assistance of a guide. The tactile handle consists of an array of vibrotactile actuators, some proximity sensors, and an embedded microcontroller. The three main parameters used to encode the information are: (1) the location of the tactile feedback, where each row of actuators corresponds to a distance (e.g., 1 ft), while each column represents a different direction (2) the intensity of the feedback, where an increase in the vibration intensity means that the user is getting closer to the obstacle, and vice versa, and (3) the timing of the feedback, in which a continuous vibration feedback can become uncomfortable to the user. In response to the last point, in the case of a continuous obstacle, e.g., a wall, the feedback is divided into given intervals of comfortable length and frequency. An ultrasound sensor measures the distance between the subject and an obstacle and sends the calculated value to the microcontroller. The microcontroller evaluates the distance and gives orders to the vibration motors to actuate. The signal coming from the microcontroller to the tactile actuator is a pulse-width modulation (PWM) signal [49]. A similar system using another mobile device is described in [60].

2.4 Tele-Robotics and Tele-Operation

Tele-operators are remote-controlled robotic devices. The first electrically actuated tele-operators were built in the 1950s at the Argonne National Lab, by Dr. Raymond C. Goertz, to remotely handle radioactive substances. When such devices are simulated using a computer, as they are in operator training devices, it is useful to provide the force feedback that would be felt in actual operations. In such cases, when contact forces are reproduced for the operator, it is called "haptic tele-operation". Given that the objects being manipulated do not exist in a physical sense, the forces are generated using haptic operator controls. Data representing touch sensations may be saved or played back using haptic technologies.

Since then, the use of haptics, particularly force feedback, has become more widespread in all kinds of tele-operators, such as underwater exploration, assembly and manufacturing, and micro-assembly. For instance, Glencross and her colleagues have developed a system called Distributed Interactive VIrtual PROtotyping (DIVIPRO) for virtual assembly and maintenance tasks [131]. The system enables collaborative and cooperative engineering designs between geographically distinct design teams. Four types of behaviors are considered: collision detection, geometric constraints, flexible pipe object simulation, and force feedback. Another example is the Collaborative Haptic Assembly Simulator (CHAS) [181]. The system is comprised of two components: the Assembly simulator (AS) and the Haptic assembly simulator (HAS). The AS allows interobject collision detection and automatic recognition of assembly constraints between the grasped component and the rest of the mechanical assembly. The HAS enables the user to touch the assembly components using a haptic device in different user actions or interaction modes: touch, grasp, move, collide, assemble, and disassemble.

As far as microassembly applications are concerned, several prototypes have been built to demonstrate the improvement that haptic information brings to micro-scale manufacturing (micro-nano and bio-manipulation). In addition to geometric scaling, with haptics it becomes possible to perform force scaling and to include complex physical models into the control loop to achieve a higher level of manipulation precision and better understanding of the environment. For example, Hollis and Salcudean [171] presents a foundation platform that can be used to perform tele-operated microassembly. The major contribution was the combination of the Magnetic Levitation Haptic Interface with a mini-factory, both developed at Carnegie Mellon University (CMU). The experimental results showed the advantages of the magnetic levitation device characterized by low levitated masses, force independence with position, and force linearity according to Lorentz[6] levitation principle. However, the main disadvantage is the low force output of Lorentz (magnetic) actuators compared to Maxwell (electric) actuators. In addition, van Strijp et al. [361] developed and tested a virtual micro world for micro-assembly tasks using a 3D display and the PHANToM Omni haptic device. The authors found that adhesive forces (not volumetric forces) determine and dominate the interaction within microassembly. A linear scaling was used to map the micro forces to the forces exerted on the user of a haptic device.

2.4.1 Tele-Surgery

In addition to problems already associated with surgical simulations, tele-surgery involves two additional issues: the coherency of the virtual scenes among all participating users, and force feedback stability when haptic information is sent over nondedicated channels such as the Internet, where there is, among other things, latency and jitter. A tele-surgery system comprises three components: (1) a master console with input devices that is known as the surgeon side, (2) a communication channel for bilateral control, and (3) a slave robot at the patient side.

Gosselin et al. [138] developed a new force feedback master arm that pays particular attention to precise manipulation and transparent behavior. The proposed system provides two-hand manipulation (i.e., the control of two slave devices simultaneously) and a dexterous manipulation representation of movements as in open surgery. Furthermore, the input devices are statically balanced to avoid involuntary movements so that a high level of safety is guaranteed.

A collaborative haptic simulation architecture for tracheotomy surgery has been proposed in [409]. Two application scenarios were considered: two doctors at geographically different places collaborating to perform a surgery; and a trainer who coaches the trainee on how to perform a surgery in a "tele-mentory" manner. The authors claim that the use of a haptic real-time controller (HRTC) guarantees a stable haptic control loop and can compensate for network delays.

At the University of Ottawa, a group of researchers developed a hapto-visual eye cataract surgery training application [107, 146]. The application supports three scenarios: (1) an instructor and a trainee – in distinct physical locations – interacting in real-time in a tele-mentor fashion, (2) a trainee learning the surgical procedure by means of perceptual cues, and (3) a trainee performing the surgery without any guidance. The developed application utilizes the CANARIE network to ensure

---

[6]In physics, the Lorentz force is the force on a point charge due to electromagnetic fields:source Wikipedia.

smoothness and transparency of the remote components. CANARIE–Canadian Network for the Advancement of Research, Industry, and Education, and its Light-path is an ongoing program allowing researchers to request and obtain dedicated CANARIE network infrastructure resources to build their own networks. Thus, a Lightpath is a dedicated high bandwidth communication channel or virtual circuit, or the concatenation of several sections of these to form an end-to-end Lightpath, providing effective bandwidth over great geographical distances [59].

2.4.2 Military Applications

There are numerous applications in military training and simulations that can benefit from the adoption of haptic technology. There are circumstances in which haptics can be a useful substitute information source, where vision and sound are not available, or are apprehensible. For instance, battlefield conditions, such as the presence of artillery fire or smoke, might mask sound and vision modalities, thus making haptics an efficient communication channel. Additionally, haptics could function as an assistive information source to sound or vision that helps the military of the future in its battle against new and hard-to-find enemies.

Kron and Schmidth [215] proposed military applications that include land mine or bomb detection and removal, collaborative training environments, casualty evacuation, and battlefield surgery. They implemented a tele-presence system, which enables an expert to execute a given task from a central control station located in a safe remote environment. It is possible that expert care could be delivered by a haptic-enabled in-field robot. In this type of tele-presence system, an expert is able to safely perform the defined task from a central control room out of harm's way.

2.5 Media

The incorporation of haptics with audiovisual media dates back to 1959, when tactile stimulation was used to enhance the movie "The Tingler" by attaching vibrating devices to the theater seats. In the 1970s, a sound speaker system, named "Sensurround", used subsonic rumbles in order to enable the audience to feel the theater shake [384]. The vibrating devices were synchronized with sound effects to enhance the audio sensory experience. Although these systems do not provide various sophisticated feelings synchronized with objects in a scene, they provide simple vibration cues that help viewers become more immersed in audio-visual media. Another example is the Showscan simulator that moves, tilts, and shakes seats in an auditorium in synchrony with the audiovisual contents displayed on a large-format screen. The digital multimedia age is moving rapidly to reach homeowners so that they can enjoy and be immersed in high quality video and audio media. High definition (HD) video and video on demand (VOD) are pushing viewers to be interested in more interactive scenarios, such as touching and manipulating video broadcasting media [71]. Lately, research indicates the greater feasibility of applying haptics into audiovisual media and proposes haptic interaction scenarios for broadcasting content such as "The TouchTV Project" [271]. In this approach, the authors focus on disconnecting the link between the audio channel of the media content and the haptic display in order to define a particular haptic channel. With this dedicated channel, they plan to distinguish which content can be created off-line from that which can be gathered and transmitted in real time. Based on this goal, they proposed two content scenarios: authored content and real-time content. In authored content, viewers are able to interact with and influence the presentation of pre-recorded content. In real-time content, the acquisition and display of haptic content occurs in real time. The authors built two systems for measuring and transmitting accelerations of a vehicle in racing applications and collision impacts of a ball in a soccer game as a demonstration. In an extended study, they adapted a force feedback gaming joystick to resemble a TV remote control in order for users to place their fingers over the actuators in a traditional way [272]. Through a series of cartoon sequences, they have investigated the application of haptics in the broadcasting media.

2.5.1 Haptic Broadcasting

Cha et al. [72] showed interest in haptic broadcasting by proposing a home shopping environment. The authors demonstrated a home shopping scenario where viewers could haptically explore 3D items such as wristwatch models. In their implementation, a 3D product model was overlaid onto a 2D video captured through a web camera. In the system, a fiducial marker position and orientation from the camera are calculated by using the Augmented Reality technique, and the 3D model is overlaid onto the marker's position with the appropriate orientation. As a result, the 3D model looks seamlessly attached to the captured scene. Afterward, viewers can touch the 3D model through a haptic device.

Gaw et al. [125] proposed an authoring environment for embedding haptic information in a video stream. Their idea is to use one graphical interface for recording haptic information and another one for playing it back. The purpose is to haptically annotate movies to allow users to feel what is happening on the screen. Additionally, Yamaguchi et al. [399] proposed a system that generates haptic feedback automatically from 2D graphics by relying on metadata that describes the movement characteristics of the media contents. Viewers can feel the motion of the objects using a 2 DOF force feedback device. The intent to enable users to interact haptically with a video stream has been proposed in [69]. They developed a framework for a haptically enabled broadcasting system that allows media acquisition, creation, authoring, transmission, viewing, and interaction based on the MPEG-4 framework. The media contains haptic properties and motion data that are physically and spatiotemporally synchronized with audiovisual media. The audio-visual media, the haptic media, and the scene descriptors are compressed with separate encoders and multiplexed into a stream that is saved in MP4 format (designed for MPEG-4 media information). This file is transmitted to viewers via a streaming server through satellite, airwaves, the Internet, etc.

Another interesting application that augments synchronous haptic feedback to video contents is presented in [297]. The application streams a YouTube video onto the local machine and presents the video to the user via a tactile player called the arm band device. The YouTube video is annotated with tactile feedback using XML notation and time stamps that specify when the tactile actuation is triggered. The application is composed of a client browser, implemented using Java-based SWT components provided with the IBM Eclipse tool, and the tactile arm band device. The haptic rendering logic is embedded in the client browser, and a Bluetooth communication module is used to connect the arm band to the computer. The arm band device is embedded with vibrotactile motors that generate vibrations at controllable amplitudes, frequencies, and durations to simulate different tactile feedback.

2.5.2 E-Commerce

Force feedback can allow a consumer to physically interact with a virtual product before purchasing it. Human hands are able to test a product by feeling the warm/cold, soft/hard, smooth/rough, and light/heavy properties of surfaces and textures that compose a product. Consumers usually like to touch certain products, such as bed linens and clothes, before buying them.

Surprisingly, little work has been completed in the field of haptic-enabled e-commerce. For instance, Shen et al. [340] proposed a scenario for the online experience of buying a car. A virtual car showroom is created, along with avatars for both the customer and the salesperson so that they can communicate in real time. Furthermore, the customer avatar can perform haptic-based functions inside the car, such as turning the ignition and the sound system on or off. The same scenario was developed in [108] within a generic framework called Unison. The framework serves to standardize the development of hapto-visual applications by providing a fixed set of services regardless of the choice of graphic or haptic software and hardware.

Another e-commerce application is introduced in [77] and aims to provide a more realistic interaction through a computer mouse system. The authors present a scenario where a customer logs onto a virtual sweater shop website and clicks on their favorite fabric. The gesture information associated with the fabric is downloaded and displayed on the local computer via the haptic mouse system. In this application, there is no need for real-time interaction, yet the correctness of the haptic modeling is still an open issue. The HAPTEX project (HAPtic sensing of the virtual TEXtiles) ushered in new avenues for research by enabling a user to perceive, touch and manipulate textiles. The goal of this project was to design an interface that enables realistic rendering of textiles and to synchronize multiple sensory feedback (haptics and visual) [239].

2.5.3 Video Games

According to Nilsen et al. [265], the gaming experience comprises physical, mental, social, and emotional aspects. We argue that, in particular, force feedback technology enhances the physical aspects of the gaming experience by providing a deeper physical reality when playing a game, improving the physical skills of the players, and allowing players to imitate the usage of physical artifacts. It is the physical aspects of the game that force feedback technology (haptics) enhances by creating a more realistic physical feeling when playing a game. Currently, a diverse spectrum of games available in the market take advantage of the force feedback effects offered by mainstream haptic interfaces.

One of the first works on the development of joystick-like haptic devices was carried out at Massachussetts Institute of Technology (MIT) and University of North Carolina at Chapel Hill (UNC), which resulted in a 3-DOF device that simulates an object's inertia and surface texture [259]. Using this device, Ouhyoung et al. [277] designed a game-like flight simulator that creates vibrations whenever the aircraft is attacked by the enemy or reaction forces on the handle whenever one fires a weapon. Currently, vibration feedback joysticks and steering wheels from companies such as Logitech are widely used as input devices in video games. In this sense, haptic research has introduced new forms of complexity in the development of games by emulating the user experience based on this particular bidirectional feedback. Pioneering attempts at introducing modern haptics to gaming include Haptic Battle Pong [263], a pong clone with force-feedback that haptically displays contact between a ball and a paddle using the PHANToM Omni device [335]. The PHANToM is used to position and orient the paddle and to render the contact between the ball and the paddle as physical forces. Haptic Arkanoid is another ball-and-paddle game where a player uses a paddle to deflect a ball so that it hits the surface of a brick wall and generates the physical impact feeling of the rebound [114]. It has been shown that playing the haptic version is more fun even though the vibration feedback is not realistic.

Nilsson and Aamisepp [266] worked on the integration of haptics into a 3D game engine. They have investigated the possibility of adding haptic hardware support to Crystal Space, an open source 3D game engine. Haptic support is being added via a plugin to the existing visual game engine with some limitations [376].

By using existing, well-developed game engine components such as Unity 3D, a scene graph library and physics engine, and augmenting them with the design and implementation of haptic rendering algorithms, it is possible to create a highly useful haptic game development environment. This can result in a rich environment, which provides players or users with a higher sense of immersion, as well as new and interesting ways to interact with the game environment [18]. In addition, this simulated world can be used to do research on applications such as physical rehabilitation, driver training simulations, and more.

There is also a haptic device called HandJive designed for interpersonal entertainment [118]. The concept is described as a handheld object that fits in one hand and allows remote play through haptic input and output. It communicates wirelessly with similar devices and provides haptic stimuli. In fact, haptic devices are becoming more accessible to the average computer and console user and will play an important role in providing innovative forms of entertainment. As further evidence, in 2008, Novint Technologies introduced the Novint Falcon device, which is affordable, even for mainstream consumers [267]. This device is now integrated with several popular video games.

2.5.4 Arts and Design

Adding force feedback to virtual sculpturing is a natural evolution that would enhance the immersion of the user and the perception of artwork. Virtual haptic sculpting, based on the constructive volume methodology, has been developed by Chen and Sun [74] to perform melting, burning, stamping, painting, constructing, and peeling interactions in real time. The authors in [89] had the vision that by using haptics in a virtual design environment, designers would be able to feel and deform the objects in a much more natural 3D setting. A sculpting system was proposed that would allow users to interactively feel the physically realistic presence of virtual B-Spline objects with force feedback throughout the design process. Blanch et al. [40] proposed a solution for reducing the classical problems of instabilities during the interaction with virtual sculptures by allowing touching and/or editing of the artwork. Thus, as expressed by Mr. Blanch, "this technique enforces the interactivity of the task and leads to an enhanced nontactorealistic feedback that increases the usability of the sculpture tool" [40]. The nontactorealistic feedback provides an expressive force feedback and refers to the use of forces computed with a psychophysical model rather a physical one, thereby changing the haptic representation of the object being manipulated.

Virtual Clay is another example of using haptics to enhance the functionality of deformable models in a natural and intuitive way [249]. The idea is that there is a natural connection between the haptic and the dynamic models; both depend on real-world physical laws to drive the realistic simulation and interaction of dynamic objects. Therefore, the Virtual Clay was designed based on dynamic subdivision-based solids models that respond to applied forces in a natural and predictive manner and give the user the illusion of manipulating semi-elastic virtual clay.

In the art of painting, DAB is an interactive haptic painting interface that uses a deformable 3D brush model to give users natural control of complex brush strokes [30]. It was found that force feedback enhances the sense of realism and provides tactile cues that help users in handling the paintbrush in a more sophisticated manner. The physical feeling of digital painting, derived from the Japanese traditional streaming art of Sumi-Nagashi, has been developed in [403]. Another haptic device, called the Haptic Desktop System, is used in drawing tasks and acts as a virtual guide through its force feedback capabilities [294].

Haptic technology has significant benefits for virtual museums [51]. It makes very fragile objects available to scholars, allows remote visitors to feel objects at a distance, lets visually impaired people feel the exhibits, and allows museums to display a range of artifacts without taking up museum space. Bergamasco and his colleagues [35] are creating the architecture of the "Museum of Pure Form" virtual reality system. Two proposed approaches have been developed: (1) placing the system inside several museums and art galleries in a network of European cultural institutions that is made available to visitors to the institutions, and (2) placing and testing the system inside a CAVE (Cave Automatic Virtual Environment: a room-sized cube whose walls are used for displaying an immersive virtual reality environment through projectors).

2.6 Mobile Haptics

With the development of touchscreen-based phones, such as the Motorola A1200, HTC Diamond, Sony Ericson w960i, Samsung Ultra Smart F520, Apple iPhone, Android T-Gl, Nokia XpressMusic 5800, and Blackberry Storm, new user interaction techniques have appeared through the use of fingers or a stylus pen. These types of mobile devices utilize gesture as the means for data input to allow much easier user interaction; for instance, users are flicking their fingers across the touchscreen for browsing through web pages or using a pinching motion for zooming into photos. Several user preference studies [382, 391] on mobile games also showed significant increases in user satisfaction for the overall game experience with advanced haptic interfaces as compared to simple on-off vibration or merely audio feedback (15–17% increase in [382]. In addition, Wei et al. [391] presented a study of replacing buttons with pen gestures in a mobile first person shooter game. The study indicated that touchscreen-based devices provide much more freedom to the users in terms of control as compared to button-based mobile phones, and eventually increased playability and the overall experience. Tactile haptic feedback is becoming common in smart phones and mobile handheld devices. Smartphone manufacturers such as Apple, Nokia, LG and Motorola are including different types of haptic technologies in their devices. In most cases, this takes the form of a vibration response to touch. Another leading edge application involves bringing the sense of touch to social interpersonal interactions between mobile phone users. Haptic information is particularly significant in social interaction [318]. A short touch can elicit strong emotional experiences, such as the comforting experience when touched by one's spouse or the feeling of anxiety when touched by stranger. Haptic stimuli can be communicated over haptic-enabled mobile phones through the use of the Hapticon Message Service (HMS), which is analogous to SMS or MMS. Hapticons are small programmed force patterns that can be used to communicate a basic 'feeling' notion through symbolic touch. Researchers have shown that humans are capable of distinguishing between seven to ten vibration patterns through the sense of touch alone, and with very little training [23]. The first step to creating Hapticons is to transform the social/emotional and physical signals into a temporal sequence of pulses of vibrations or simple vibrotactile patterns. This can be achieved by changing one or more basic parameters of the vibratory patterns such as frequency, amplitude, waveform, and duration for each Hapticon. These vibratory patterns can be stored on the mobile devices and played when a specific corresponding signal is sent from the other communicating party. There has been ongoing research into the design of the Hapticons so that they correspond to different emotional images. For instance, in [91], nine vibrotactile patterns were designed to represent nine emotional images. The authors showed that, for example, two different vibration patterns with different duration must be combined to mimic the feeling of crying.

Currently, researchers are fostering their interest in incorporating the sense of touch to facilitate social and interpersonal communication systems [105]. Haptics is crucial for interpersonal communication as a means to express affection, intention, or emotion. Examples are a handshake, a hug, or regular physical contact [50]. Several studies have confirmed that infants deprived of skin contact lose weight and may even become ill [262]. Furthermore, studies of human infants reveal that the absence of affectionate touch can cause social problems [139]. There have been significant efforts to incorporate haptic modality into interpersonal communication systems.

One of the earlier systems for interpersonal communication is the TapTap prototype [45], which is a wearable haptic system that allows nurturing human touch to be recorded, broadcast, and played back for emotional therapy. The tactile data is transmitted asynchronously. Another commercially available product is the Hug Shirt^{TM} that enables people to feel hugs over distance. It is described in [88]. The shirt is embedded with sensors and actuators to read or recreate the sensation of touch, skin warmth, and emotion of a hug (heartbeat rate), sent by a distant lover. The hugger sends hugs using a Java-enabled mobile phone application. An SMS is sent through the mobile network to the loved one's mobile phone, which then delivers the hug message to the shirt via Bluetooth. Another tele-haptic system that enables interpersonal interactions is described in [368]. In this system, an Internet pajama is developed to promote physical closeness between a child and their remote parent. The pajama reproduces hugging sensations that the parent applies to a doll or teddy bear in place of the child.

An interesting research project has been performed at the MCRlab of the University of Ottawa. They developed the KissME system, which is a neck gaiter that enables people to receive kisses or haptic touches over distance [296]. Embedded in the gaiter are actuators that recreate the emotion of the kiss from the remote user. A Bluetooth connection was used for the communication between a mobile phone and the gaiter in order to send a "kiss" message from one user to another. Furthermore, the KissMe system was used to bridge real and virtual words by converting a kiss in the virtual world (between two avatars in Second Life) to a physical kiss using a neck piece [296]. The system later evolved into what the team called the HugMe system [105]. The HugMe system simulates haptic interpersonal communication between two users. The remote person is wearing a

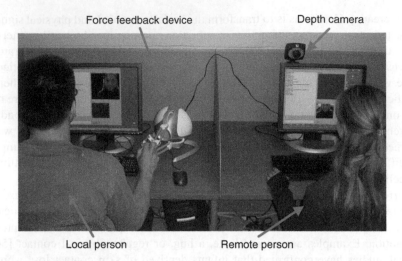

Fig. 2.3 The HugMe system

haptic suit (haptic jacket) that is capable of simulating a nurturing touch. The local person uses a haptic device to communicate his/her feelings to the remote person. A depth camera (2.5D camera) is used to capture the image and depth information of the remote person and send it back to the local person's computer. The local person can touch the video contents with the force feedback device while the remote person receives synchronous touch via the haptic jacket (Fig. 2.3).

2.7   Haptics and Virtual reality

The possibilities of integrating haptic interactions with Linden Lab's multiuser online virtual world Second Life [234] are investigated by Pascale et al. [283]. Once connected to Second Life, the users can view their avatars in a computer simulated 3D environment, and they can participate in real-time task-based games, play animations, and communicate with other avatars through instant messaging and voice. The social communication aspect of Second Life is hugely popular, and the community has millions of users. Moreover, its open source viewer provides a unique opportunity to extend it further and equip it with the haptic interaction modality [352]. Another interesting effort to integrate haptic interactions in Second Life is presented in [174, 297]. The Second Life haptic interaction prototype system attempts to bridge the gap between virtual and real world events by incorporating an interpersonal haptic communication system in Second Life. The developed system works as an add-on and is loosely coupled to the Second Life viewer. The haptic and animation data are annotated in the virtual 3D avatar body parts. The 3D avatar and the annotated body parts represent a real user who receives input through gesture, mouse, speech, or text. This produces emotional feedback such as touch, tickle, and hug to the real user, through a previously developed haptic jacket system [66] that is composed of an array of vibrotactile actuators. The haptic jacket provides the funneling illusion based touch haptic feedback. The funneling illusion describes a phantom sensation phenomenon midway between two stimulators (e.g., vibrotactile or sound stimuli), where they are presented simultaneously at adjacent locations [15, 32].

2.8   Education and Learning

There has been a growing interest in developing haptic interfaces that allow people to access and learn information in virtual environments. A system for constructing a haptic model of a mathematical function using the PHANToM device was introduced and partially implemented in [332]. The program accepts a mathematical function with one variable as input and constructs a haptic model made of balsa wood with the trace of the function carved into its surface. This prototype has been extended in [333] to display functions of two variables, introduce sonification of functions of one variable, and improve the user interface.

A virtual reality application combined with haptic feedback for geometry education has been recently investigated [265]. The proposed system presents a haptic 3D representation of a geometry problem's construction and solution. The performance evaluation showed that the system is user friendly and has provided a more efficient learning approach. Also, a multimedia system that incorporates visual, audio, and force feedback has been developed in [242]. Their preliminary results demonstrated that adding force feedback can enhance the learning process, especially in languages that are based on non-Latin characters such as Arabic, Japanese, or Chinese. This application guides users, allowing them to see, hear, and feel the character's shape. Their approach uses a Dynamic Time Wrapping technique to recognize the character that has been evaluated in its interface. Their multimedia learning tool supports five languages: Arabic, English, Chinese, Japanese, and French, and it is implemented under the PHANToM Omni haptic device.

Another application, which simulates a catapult, has been developed to enable users to interact with and learn about the laws of physics by utilizing a force feedback slider (FFS) interface [213]. The FFS is a motorized potentiometer limited to 1-D of movement (push/pull along a line). The user simply grabs the slider and moves the handle. It is claimed that the force feedback helps users in creating a mental model to understand the laws of physics.

2.9   Haptics for Security

Using haptics as a mechanism for identifying and verifying the authenticity of users is a novel avenue of haptic research. The feasibility of the haptic biometrics approach has already been proven by Orozco et al. [110, 273, 274].

A haptic-biometric system has been proposed in which physical attributes such as position, velocity, force, and torque data are extracted from the interaction of the haptic end effector within a virtual environment. Thus, data generated through a user performing a specific task, such as signing a virtual check, were continuously measured and stored. Subsequently, the proposed haptic system generated a biometric signature from the measurement and evaluation of that specific data, which was used for authentication purposes.

Similar work in this domain includes the design and implementation of a graphical password system that incorporates the sense of touch using haptic interfaces [275]. The system utilizes the physical attributes captured during human–computer interaction, including, for example, pressure and velocity, and uses them as 'hidden' features to increase the resiliency of the system. The authors claim that the proposed system is more resistant to well-known security system flaws other than previously known graphical password schemes. Using haptics, the proposed system was able to integrate pressure as a binary input during the generation of a graphical password. This increased the resiliency of such a scheme to both dictionary attacks and shoulder surfing attacks. Figure 2.4 shows both 5 × 5 and 8 × 8 grids used to draw a graphical password. This system has a very low likelihood of a biometric security system incorrectly accepting an access attempt by an unauthorized user (False Acceptance Rate) or rejecting an access attempt by an authorized user (False Rejection Rate).

Another approach, which is based on a multilayer perception(MLP) neural network, was adopted to identify a user by analyzing a handwritten signature and its associated haptic information, such as pressure [16]. In their approach, a handwriting environment provided a virtual scenario where users could write their signature on a virtual plate. The rich haptic information, such as force, velocity, and angular rotation, were gathered as the key elements to identifying users who took part in their experimental work.

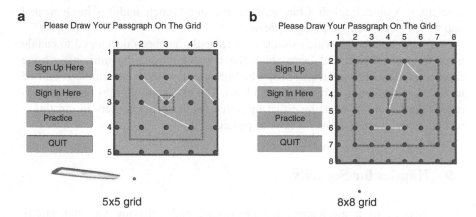

Fig. 2.4 The graphical password with haptic interaction

2.10 Closing Remarks

This chapter covers both historically significant and recent work relevant to haptic technologies and applications. It is worth mentioning that even with the recent significant progress in haptic technologies, the incorporation of haptics into virtual environments is still in its infancy. A wide range of human activities, including communication, education, art, entertainment, commerce, and science, would forever change if we learned how to capture, manipulate, and create haptic sensory stimuli that are nearly indistinguishable from reality. For the field to move beyond what is considered to be state of the art today, many commercial and technological barriers need to be surmounted. First, business models and frameworks are needed to make haptic devices practical, inexpensive, and widely accessible, with the ultimate goal being to have haptic devices used as easily as the common computer mouse.

Chapter 3
Human Haptic Perception

3.1 Introduction

In this chapter, we discuss the tactile and kinesthetic perceptual processes and the functions of the human perceptual system. An understanding of the human perception process is essential to the design and development of haptic devices and software, especially in order to maximize their performance and cost efficiency. For example, the Pacinian corpuscle, one of the four major types of mechanoreceptors in human skin, detects rapid vibrations of about 200–300 Hz; therefore, to stimulate these receptors, the vibration range of motors used in vibrotactile devices do not need to operate at frequencies over 300 Hz. Additionally, this chapter introduces haptic perceptual illusions. These can play an important role in fooling human haptic perception and generating a more complex touch sensation than the stimulus actually delivers.

3.2 Touch and Cognition

Touch is different from other senses in that it consists of a closed loop and a bidirectional channel of both sensing and acting. It depends on physical contact, and its receptors are spread over the entire body. In fact, touch relies on action or exploration to stimulate perception, which can be either passive or active. Passive tactile perception (cutaneous perception) is limited to the zone of contact with objects. While specific discriminations are still possible, tactile perceptual capacity is limited due to the lack of any exploratory movements. Nonetheless, in order to understand a given object, voluntary movements must be made to compensate for the smallness of the tactile perceptual field. The resulting kinesthetic perceptions are essentially linked to the cutaneous perceptions generated by skin contact to form the tactile-kinesthetic action, or active touch. In audio or visual signals such as speech, music, or an image, the order of the sequence of stimuli carries a meaning.

A. El Saddik et al., *Haptics Technologies*, Springer Series on Touch and Haptic Systems,
DOI 10.1007/978-3-642-22658-8_3, © Springer-Verlag Berlin Heidelberg 2011

In contrast, touch can perceive stimuli in any order and then mentally build up a whole picture even while the eyes explore a wide scene or a large picture. Consequently, touch provides information about spatial and physical properties of the environment, such as texture, mass, direction, distance, shape, size, etc. Haptic stimulation can be generated through heat, vibration, and pressure by applying several forces, which possibly results in a kind of skin deformation. We distinguish between three main types of haptic receptors: (1) thermo receptors, which are receptors for perceiving the temperature that signals heat or cold information; (2) nociceptors, which are sensory receptors responsible for the perception of pain; and (3) mechanoreceptors, which respond to mechanical actions such as force, vibration, and pressure. The first two types are considered cutaneous receptors while the third type of receptor can be found in skin, muscle tendons, and joints.

The areas of the skin that are mobile, deformable, and contain a dense collection of sensory receptors are the most effective in terms of tactile perception. These areas include the areas around and inside the mouth and those in the arm-hand system. From a cognitive perspective, the latter constitutes the real haptic perceptual system, acting on the environment and perceiving stimuli from the environment at the same time. Moreover, the hands are the motor organs that are used in reaching, holding, transporting, and transforming objects in our daily lives. In the next section, the human haptic system is more precisely introduced with special focus on the arm-hand system.

3.3 Human Haptic System

The human haptic system consists of four components: namely, the mechanical, sensory, motor, and cognitive components. The mechanical component of most significance is essentially the arm-hand system. This component consists of the upper arm, the forearm, and the hand, which, as a whole, possesses more than twenty-eight degrees of freedom for dexterous exploration and manipulation. The sensory (or somesthetic) system includes large numbers of various classes of receptors and nerve endings in the skin, joints, tendons, and muscles. Typically, a physical stimulus activates these receptors and causes them to convey sensory information (mechanical, thermal, pain, etc.) of the touched object via the afferent neurons to the central nervous system. The brain, in turn, analyzes and "perceives" this information and issues appropriate motor commands to activate the muscles and initiate hand or arm movements. This happens through the efferent nerves, which carry nerve impulses out of the central nervous system. Figure 3.1 shows the haptic interaction system in the human body.

The human haptic system perceives two types of information: tactile or cutaneous, and kinesthetic or proprioceptive; however, these sources are not mutually exclusive and are often perceived as a combination of the two. Tactile information

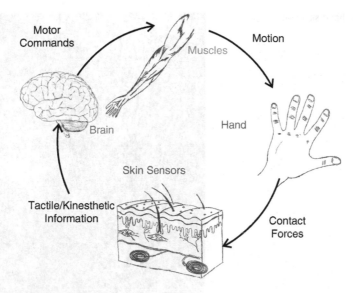

Fig. 3.1 Human haptic system

is conveyed when the human hand is passive and stationary while in contact with an object. Kinesthetic information is expressed during active and free motion of the hand. For instance, we perceive an object's shape and texture through the tactile stimulus, which is mainly provided by the tactile receptors in the skin. In order to handle, grasp, or squeeze an object, our hands must apply appropriate forces. Therefore, kinesthetic information needs to be gathered from the position and motion of the hand and arm, as well as the forces acting on them, to give a sense of the total contact forces, surface compliance, and weight [394]. Eventually, all sensing and manipulation interactions that are performed actively with the normal hand involve both types of information.

3.3.1 Mechanical Structure of the Arm-Hand Haptic System

The human arm-hand haptic structure roughly consists of a broad palm attached to the forearm by the wrist joint. Opposite to the wrist, and at the outer edge of the palm, are five digits: the thumb and four fingers. The fingers can be folded forward over the palm for holding objects. The forearm consists of the distal area of the arm between the elbow and the wrist. This human structure performs many daily tasks ranging from highly meticulous and dexterous activities to simple lifting of weight. The study and research of such a complex system has fed into the design and development of dexterous arm-hand mechanical systems for human-assisted manipulation tasks, among other applications.

Fig. 3.2 The hand skeleton
structure

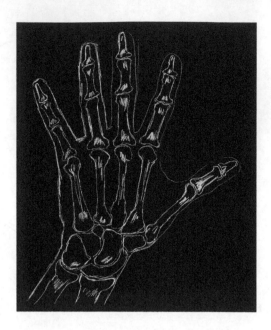

3.3.1.1   Hand Anatomy and Its Mechanical Replication

The human hand is one of the most sophisticated parts that interacts with the
environment. The hand allows us to perform fine and gross motor actions such
as displacing tiny objects and grasping. The human hand, as shown in Fig. 3.2, is
composed of many small bones called carpals, metacarpals, and phalanges. The
metacarpals articulate with the carpal bones, which, in turn, articulate with the ulna
and radius bones of the forearm to form the wrist joint [31]. Each finger has a
metacarpal bone and a proximal, middle, and distal phalanx. Exclusively, the thumb
does not have a middle phalanx. From an anatomical perspective, the human hand
has 27 bones: eight bones of the wrist are arranged in two rows of four; five bones
of the metacarpus (or palm), one for each digit; and 14 digital bones (phalanges),
two in the thumb and three in each finger. The carpal bones fit into a shallow socket
formed by the bones of the forearm.

Different electromechanical and kinematic models of the hand's finger tendons
and their relations to the finger joints have been proposed to characterize the human
hand haptic system and to replicate the hand's dexterous tasks [235, 355, 370].
These works are related to robotic hand systems, which are beyond the scope of
this book. However, we would like to briefly list some relevant works that describe
several features worth considering during the design of haptic devices. For instance,
the aspects of static and kinetic friction are major challenges in the control of a
mechanical hand model. Also, linearity relationships between tendon displacement
and joint angles play an important role in deriving accurate, dynamic models of such
mechanisms. Currently, no tactile display can present multiple cutaneous sensations

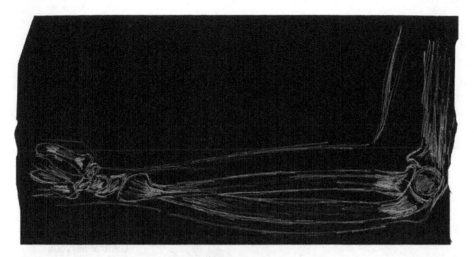

Fig. 3.3 The human forearm

at once. Although touch is a low-resolution sense and can only be applied locally, it demonstrates some superior properties to vision, which is often hampered by geometry, surface reflection, and lighting [359].

3.3.1.2 Forearm

The forearm is the structure on the upper limb that is sandwiched between the elbow and the wrist. As shown in Fig. 3.3, the bones of the elbow are the humerus (the upper arm bone), the ulna (the larger bone of the forearm located on the opposite side of the thumb), and the radius (the smaller bone of the forearm located on the same side as the thumb). The elbow behaves like a hinge joint between the humerus and ulna; the movement is along one direction and is comparable to a mechanism by which a door or a lid opens and closes. However, there is a second component to this joint where the radius (the radial head) and humerus meet. This complicates the joint because the radius has to rotate so that the hand can be turned palm up or down. At the same time, it has to slide against the end of the humerus as the elbow bends and straightens. The joint is even more complex because the radius has to slide against the ulna as it rotates the wrist as well. As a result, the end of the radius at the elbow is shaped like a smooth knob for sliding against the ulna and has a cup shape that fits onto the end of the humerus.

Since grasping is linked with the forearm anatomy, we will go over the guidelines for grasping tasks. In robotics, when a robot arm grasps an object, the interaction involves the real world. This means that objects have a size, weight, and form that all directly impact the grasp activity. Thus, the capability of grasping for a robot involves many considerations to cover even basic human activities.

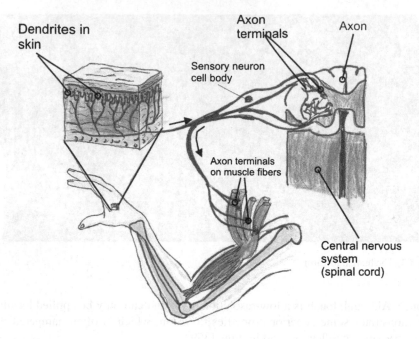

Fig. 3.4 Human sensory system

3.3.2 *Human Sensory System*

In order to understand the neural basis of perceptual processes, it is important to study the characteristics of the somesthetic system. It is the sensory system associated with the human body, including skin senses, proprioception, and the feeling in the internal organs. This system is a nonhomogenous entity since its sensory receptors are widely dispersed and are functionally diversified. This section presents an overview of the structure of the touch sensory system in humans (see Fig. 3.4). The overview is divided into three parts: (1) touch receptors, (2) connections to the brain, and (3) the touch system within the brain [290]. Touch receptors are the "input devices" of the human body that collect information through tactile and kinesthetic perceptions. This is the main source of haptic data where the various physical quantities, such as pressure and temperature, are given a common representation. The connection to the brain is the "wiring" between the brain and the receptors where a lot of preprocessing takes place. Finally, the touch system within the brain is where the information is processed into its final perceivable form.

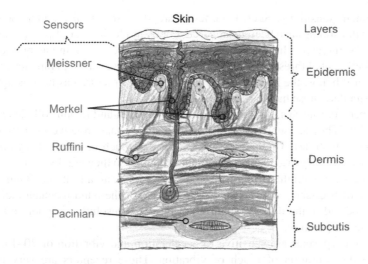

Fig. 3.5 The skin layers

3.3.2.1 Touch Receptors

The function of a sensory receptor is to transform a physical stimulus into a signal that can be processed further by other parts of the nervous system. The first step in this transformation is to cause a change in the electrical conductance of the channel protein, and thus change the conductance of the membrane channel of the receptor; this process is referred to as the chain of transduction. The second step is to create an impulsive discharge, the action of a nerve cell generating an impulse. The change in the conductance of the membrane channel causes a change in the membrane's potential, which leads to the propagation of the touch signal. In fact, second-order neurons transmit the generated signal up the spine and into the thalamus region of the brain. Here, third-order neurons complete the path to the cortex where the corresponding sensation of temperature, pain, and/or pressure is registered.

In tactile perception, information from the mechanical deformation of the skin is coded by cutaneous mechanoreceptors situated in the different layers of the skin, as shown in Fig. 3.5. The human tactile sense is composed of four kinds of sensory organs in the hairless skin: Meissner's corpuscles, Pacinian corpuscles, Merkel disks, and Ruffini endings [343]. Their sensitivities depend on their size, density, frequency range, and nerve fiber branching. These receptors have been classified based on their adaptive properties (rapid versus slow) and the characteristics of their receptive fields (small and highly localized versus large). Meissner corpuscles and Merkel disks have small, localized receptive fields, whereas Pacinian corpuscles and Ruffini endings have large and less localized receptive fields. Many functional features of cutaneous mechanoreceptors have been subject to extensive study, such as the rate of adaptation to stimuli, the location within the skin, the mean receptive areas, the spatial resolution, the response frequency rate, and the frequency for maximum sensitivity. Such parameters are at least partially understood, and important thresholds have been discovered. For example, in a set of psychophysical experiments, the capability of the human fingertip to detect strain has provided us with a specific set of threshold metrics. In addition to these sets of parameters, which are described below, the hand and forearm anatomic structure has been adapted in order to simulate those haptic sensory principles.

Meissner's corpuscles are found both in hairless skin and in hair follicle endings in hairy skin. They are sensitive to light touch, local shape, relative velocity (slip), and flutter (10–60 Hz). They elicit the sensation of touch or flutter. They are also characterized by very fast adaptability to the stimulus, so they are only active during the initial contact with the stimulus. The mean receptive area is about 13 mm^2. The proportion of Meissner's corpuscles is about 43% of the hand mechanoreceptors. They are located in the dermis (shallow). The spatial resolution is poor and is not sensitive to temperature.

Pacinian corpuscles are sensitive to acceleration and vibration of 70–1,000 Hz and elicit the sensation of touch or vibration. These receptors are very fast in adapting to the stimulus. They are found both in hairless and hairy skin below the dermis and subcutis (very deep). The mean receptive area is about 101 mm^2. The proportion of the Pacinian corpuscles is about 13% of the hand mechanoreceptors.

Merkel's disks are sensitive to small-scale shape and pressure, and they elicit the sensation of touch and pressure. They adapt slowly to stimuli, and thus are active during the entire stimulus event. From a location perspective, they reside on the dermis border near the epidermis (shallow), and they cover about 25% of the hand mechanoreceptors. Their mean receptive area is about 11 mm^2. Their response frequency range is 0.4–100 Hz.

Ruffini endings adapt slowly to skin stretch and directional force stimuli. From a location perspective, they reside in the dermis (deep) and they cover about 19% of the hand mechanoreceptors. Their mean receptive area is about 59 mm^2. Like Merkel disks, the response frequency range is 0.4–100 Hz.

Hairy skin receptors are characterized by a special low resolution, indicating that they do not effectively perceive a specific geometric structure of an object. Consequently, actuators in a haptic device for texture perception must be applied to hairless skin areas (such as on the palms and fingertips), while those conveying vibratory information can be activated anywhere on the body [144].

The receptors that support the kinesthetic sense are classified into four categories, two in the joints and two in the muscles: Golgi-type endings in joint ligaments (joint torque), Ruffini endings contained in the joint capsules, Golgi tendons monitoring muscle tension, and muscle spindles contained in the muscles to measure static position and movement [144]. Working together, these receptors provide information about joint angles and muscle length, tension, and rates of change. They provide information on the movement of joints, the movement's velocity, and the contractile state of the muscles controlling the joint. Combined with the information from the motor and cognitive systems, this produces the perceived limb position and movement. It is worth mentioning that the force control and perceptual bandwidth differs from person to person. For instance, the maximum frequency with which a typical hand can command motion varies from 5 to 10 Hz, while the position and force signal bandwidths range from 20 to 30 Hz [54].

3.3.2.2 Connections to the Brain

Sensory information coded by cutaneous and proprioceptive receptors is transmitted to the central nervous system by two separate, major ascending pathways: the dorsal column-medial lemniscal system and the anterolateral (or extralemniscal) system. Of these two pathways, only the dorsal column-medial lemniscal system is of significant importance to tactile and kinesthetic perceptions because it transmits information involved in cutaneous and proprioceptive sensitivity rapidly (from 30 to $110 \, \mathrm{m \, s^{-1}}$).

Two different types of nerve fibers convey signaling from receptors to the spinal cord: first-order neurons and second-order neurons. These are connected through synapses in either a many-to-many or a many-to-one fashion. The first-order neurons are physically divided into two groups/systems: the spinothalamic system and the lemniscal system. The spinothalamic system is a bundle of neurons that transmits sensations of temperature and pain, whereas the lemniscal system comprises the mechanoreceptors. The second-order neurons interact with neurons either leading to the brain or down to glands and muscles, thus giving rise to some reflexes.

3.3.2.3 Touch System in the Brain

The sensory information from the skin receptors is carried to a layer on the surface of the brain called the somatosensory cortex. This outer layer of the brain is about one quarter of an inch thick. The mapping of the human body on the cortex is known as the "homunculus". The body parts that have a high acuity of perception, such as the lips and fingers, comprise large areas of the homunculus, and less acute parts comprise much smaller areas. The homunculus is located in the primary somatosensory cortex (SI). However, somatosensory information is also processed in the secondary somatosensory cortex (SII), as shown in Fig. 3.6.

From the somatosensory cortex, messages about sensory input are sent to other areas of the brain, such as motor areas for generating actions.

3.3.3 The Motor System

The motor subsystem comprises contractile organs (such as muscles) by which movements of the various organs and body parts are affected. The median and ulnar nerves are the major nerves of the hand. They spread along the length of the arm and transmit electrical impulses to and from the brain, generating sensations and motion. The movements of the human hand are accomplished by two sets of muscles

Fig. 3.6 Sensory input
mapping onto specific brain
areas

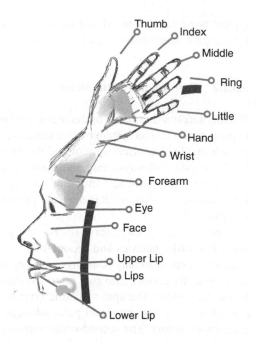

and tendons: the flexors, for bending the fingers and thumb, and the extensors, for straightening out the digits. The flexor muscles are located on the underside of the forearm and are attached by tendons to the phalanges of the fingers. The extensor muscles are on the back of the forearm and are similarly connected to the phalanges. The human thumb has two separate flexor muscles that move the thumb in opposition and make grasping possible.

3.3.4   Haptic Cognition

The scope of human cognition is defined by the five senses, namely touch, audio, taste, sight, and smell. Currently, there is no multimedia system that can integrate and accurately incorporate all the senses within a simulation environment. In tele-manipulation systems, the main idea is not only to allow users to control objects within a simulation, but to control objects in the real world as well. In order to tele-manipulate objects in the real world, the interaction must have at least two sensory channels or media: the vision and the haptic channels. Haptic perception focuses on understanding which parameters are involved in the process through which humans perceive the reality of touching. Significant research has been dedicated to simulating a computer-generated reality that can be manipulated through haptic devices. Current research has been focused on measuring human–computer interactions in terms of touch and sensations, commonly referred to as "haptic quality".

Haptic cognition involves tactile perception through the skin and kinesthetic perception through the movements and positions of the joints and muscles. The study of haptic perception brings together many disciplines, such as neurophysiology, psychology, and psychophysics. We divide haptic perception into two categories: haptic exploration and haptic manipulation. Haptic exploration has the primary goal of recognizing an object's properties, whereas haptic manipulation involves perceiving modifications in the environment.

3.3.5 Haptic Exploration

The physiological and spatial arrangements of tactile sensors allow humans to perceive various spatial features in parallel and assemble a representation of objects and their features. For example, applying static pressure to a surface provides significant information about hardness [358], whereas it is less informative with regards to roughness [169]. Klatzky et al. [211] found the following: lateral motion (tactile) facilitates the discrimination of texture; enclosure with the hand (kinesthetic) gives information about the volume and global shape of the object; static contact (tactile) with the object provides information about the temperature, while contour following (tactile) gives insight of the shape of the object; weight (kinesthetic) is determined by unsupported holding of the object in question; and finally, kinesthetic force (pressure) provides information about the object's hardness characteristics.

The discrimination of softness of virtual objects through haptic devices is still a great challenge for current technologies. This is because the human hand is a complex model to simulate. Ambrosi et al. [17] investigated how information has to be handled in order to distinguish an object's softness and have provided elements to be used in the design of haptic devices for practical uses. Their observations show that a finger touching the surface of a specimen at different orientations does not affect the discrimination of softness, nor is it very sensitive to the location of the contact area on the finger surface. They also consider haptic discrimination of softness as fundamentally invariant with translations and rotations of the contact area.

Experimental work has been carried out to validate a model related to a simplified form of tactile information. Five psychophysical experiments were conducted to test the state of recognition, consistency of perception, perceptual thresholds, psychometric functions, and perceptual granularity. These experiments were performed with an implementation of sensors and actuators that define the model, which they called the Contact Area Spread Rate paradigm (CARS). The CARS hypothesis argues that a large part of the necessary haptic information to discriminate an object's softness by touch is contained in the law that relates the overall contact force to the area of contact, that is, in the rate by which the contact area spreads over the finger surface as the finger pressure on the object is increased.

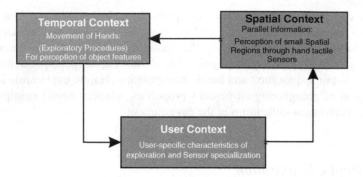

Fig. 3.7 The role of context in haptic perception

Haptic perceptual exploration allows a user to recognize objects and their spatial features in a virtual/augmented reality and is characterized by three contextual variables: spatial, temporal, and user contexts, as shown in Fig. 3.7 [198]. In contrast to visual recognition, which starts from observing the whole and then the parts, haptic exploration is a mental process that constructs the whole after the perception of the parts [305].

In the haptic perception process, the spatial position of the fingers and the palm, while in a static grasp of an object, could be modeled as the spatial context. Neurophysiology researchers have shown that different regions of the hand are specialized to perceive particular spatial features, such as texture, shape, weight, and material [193]. For instance, shape and hardness are related to a spatial distribution of stress or deflection when a human or artificial finger interacts with an object. Moreover, Emura and Tachi have found that a human finger pad has the same structure as a human eye in the sense that it has a higher density of mechanoreceptors in a small region at the center and better resolution over the central region than in the peripheries [111]. An approach based on this principle has also been proposed as a solution to haptic perception during general motion. This approach, referred to as "haptic servoing", extracts feature-based data from a visual control technique called "visual servoing". Visual servoing is based on a setup of one or two cameras and a computer vision system, whereas the haptic servoing approach is much more complex due to the elastic theory involved.

As part of haptic perception, the exploration and recognition of convex shapes through tactile sensing has also been investigated by developing internal and external volumetric approximations of the unknown object [64]. The approach of exploring a polyhedral model is based on two facts: the tactile information is naturally conceived in 3D space, and it is intrinsically thinly dispersed. Thus, the proposed technique takes advantage of an effective selection strategy based on volumetric approximation. This addresses sensing along directions where vagueness related to the explored object is greater, and it results in early pruning of incompatible objects to improve recognition performance. However, this study recognizes that complex kinematics exist in multifingered hand exploration strategies when mapping two particular convex representations. Therefore, these strategies are considered only with respect to a "single-finger" device. Afterward, Robles de la Torre et al. [310] investigated the significance of cues related to geometry and the range of forces experienced during exploration; he demonstrated that force cues provide richer content than geometric cues when discriminating the shape of an object. Therefore, static touch is a complementary task in haptic perception that provides enough information to construct a conceptual representation of an object.

When trying to comprehend a whole object, voluntary hand movements (exploratory procedures) made to compensate for the smallness of the tactile perceptual field can be modeled as the temporal context [198]. Researchers in the psychology of haptic perception have proven that perception and action are closely related in the haptic modality [305]. Various attempts have been made to study the manual exploratory procedures of blind and sighted individuals [219, 220], and based on their findings, it has been concluded that "the haptic and visual systems have distinct encoding pathways". The haptic system is more focused on discriminating object characteristics other than shape. Previously in this domain, the work of Ernst Heinrich Weber described the importance of tactile perception action in terms of object feature discrimination. From his experimental work, he concluded that "the shape and texture are not discovered by touch, unless the finger is deliberately moved over the surface of the test object" [316].

Finally, the user's style of haptic perception and their mannerisms can be modeled as the user context. Being aware of this context allows for a customization of the haptic rendering and visualization schemes to the user's style of haptic exploration and their cognitive strategy in assembling piecewise information into haptic object memory [198].

3.3.6 Exploration of Perceptual Metrics

The maximum frequency with which tactile and force stimuli can be sensed is called the sensing bandwidth, and the speed with which users can respond or act is called the manipulation bandwidth. These two bandwidths are asymmetric. In fact, the sensing bandwidth is much higher than the manipulating bandwidth, which means that humans can sense haptic stimuli much faster that they can respond to them. The bandwidth with which a human can react to unexpected force/position signals is 1–2 Hz. The ability of hands and fingers to exert force is about 5–10 Hz. The human fingers and hand require the force input signal to be present at about 30 Hz (from 20 to 50 Hz) in order to perceive meaningful information. It is also estimated that the bandwidth beyond which the human hand and fingers can discriminate two consecutive force input signals to be at about 320 Hz, after which they are just sensed as vibrations. The bandwidth of tactile sensing varies from 5 Hz up to 10 kHz.

Many studies have used the well-established metrics of roughness, hardness, and stickiness to characterize the state of haptic perception [170]. These are described below, yet there are also lesser studied metrics such as blurriness, distortion, and aberration, of which further information can be found in [113].

Roughness measures the small-scale variations in the height of a physical surface; indeed, Klatzky et al. [210] have stated that "surface roughness is particularly salient to the tactile sense". Several studies and experiments have been conducted to quantify roughness of texture elements. For instance, Cascio and Sathian found that temporal cues do indeed contribute to tactile texture perception [63]. It has been shown that the perceived roughness requires lateral movement between skin and surfaces and depends on temporal cues. Perceived roughness increases with the increment of inner-element spacing, grating groove width (G) [63, 221]. In addition, Sathian et al. [330] stated that "humans subjects scaled gratings of altering grooves and ridges for perceived roughness". In fact, Sathian et al. [330] found that roughness increased with an increase in G with an increase in ridge width (R). Cascio and Sathian also found that peripheral neural responses to gratings depend quantitatively on G and a grating temporal variable (Ft) [63]. The quality of roughness has also been researched by extending the spectrum from very fine to very coarse textures [169]. The perception of fine textures is largely attributed to vibrations set up by the relative movement of the skin and the stimulus, whereas coarse textures are perceived mainly based on their geometric properties (when temperature, hardness, etc. are held constant). Other formal studies and experimental observations have been made regarding the impact of contact areas and the relative size (or curvature) of the featured surface to the size of the contact probe (or finger) on identifying fine surface features [276].

Hardness is characterized by the resistance force that is normally applied to the surface. Many researchers, such as O'Malley and Goldfarb [270], focused their experimental work on surface stiffness and force saturation effects. Quantitative data on the effects of force saturation has yielded several conclusions. First, haptic interface hardware may be capable of conveying significant perceptual information to the user requiring only low levels of force feedback. Second, higher levels of force output may improve the simulation in terms of perceived realism. Finally, haptic interface hardware may transmit valuable perceptual information to the user if a low level of simulated surface stiffness is considered. The last conclusion was derived after an experimental study based on the three psychophysical concepts (detection, discrimination, and identification) was conducted in order to characterize the effects of virtual surface stiffness on haptic perception in a simulated environment.

Stickiness is a physical phenomenon of adhesion and cohesion that depends on the friction and relative resistance to lateral force that exists between the subject's finger and the surface [201]. Softness is an example of this tactile information that provides a good threshold in the discrimination of haptic perception. Ambrosi et al. [17] investigated the way this information is handled to distinguish softness of objects and provide practical elements in the design of haptic devices.

3.4 Concept of Illusion

There are some cases where a visual illusion creates an ambiguous situation and confuses the brain. For example, when sitting on a train and looking out the window at a neighboring train, if the other train starts moving, there is an ambiguous situation: which train is actually moving? In either case, the brain will come up with a unique – right or wrong – answer to this ambiguous situation. To resolve ambiguities, the brain uses constraints by comparing the newly acquired information with knowledge of previously experienced situations and information.

For centuries, perceptual illusions were thought to concern only the visual system due to some specific properties such as color, and temporal and spatial sensitivity. Three important geometric illusions are well discussed in the literature: the Muller–Lyer illusion, the vertical–horizontal illusion, and the Delboeuf illusion.

The Muller–Lyer illusion has frequently been studied by psychologists. The illusion consists of two identical lines that are actually perceived as being of different lengths due to the presence of "fins" with a particular orientation placed at each end of each line. The evaluation of the length of a line segment changes according to the orientation of the arrowheads situated at either end (see Fig. 3.8).

In the vertical–horizontal illusion, the length of the vertical segment is overestimated when it is compared with the same segment in a horizontal orientation (see Fig. 3.9).

In the Delboeuf illusion, the perception of the size of a circle changes if it is inserted into a larger concentric circle, as shown in Fig. 3.10.

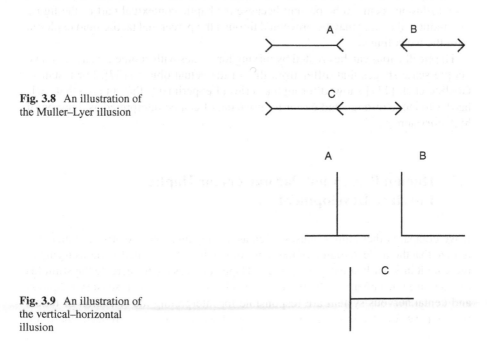

Fig. 3.8 An illustration of the Muller–Lyer illusion

Fig. 3.9 An illustration of the vertical–horizontal illusion

Fig. 3.10 An illustration of the Delboeuf illusion. The inner circle (B) is perceived as larger than the identical circle (C) because it is inserted in an exterior concentric circle (A)

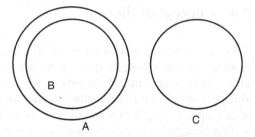

For many years, the Muller–Lyer illusion was studied as a purely "visual" illusion, but like many other optical illusions, evidence now shows that illusions also occur when stimuli are presented tactually (for review, see [100, 126]). Therefore, it has been established that perceptual illusions occur in all sensory modalities even though the most widely known and studied are visual illusions. The study of haptic illusions is still in its infancy.

In a study presented by Gentaz and Hatwell, the presence of the Muller–Lyer illusion in both vision and haptics seems to have similar results. As a common factor, the bisection observed in the horizontal and vertical illusion studies affects the visual and the haptic modality in the same way. However, contrary to the Muller–Lyer illusion, there are also factors specific to each system. In particular, the exploratory movements affect only the haptic illusion in the haptic modality, including its presence in those who have early blindness, and invalidates the exclusively visual explanations of this error. Also, they discussed that in the Delboeuf illusion only the visual illusion seems to be present because the haptic contextual part of the figure that induces the error may be prevented through the perceptual tactile field explored with the index finger.

Haptic illusions can be created by mixing force cues with geometric cues to make people sense shapes that differ from that of the actual object [137]. For instance, Gosline et al. [137] showed through a series of experiments that an area that feels harder to move through and easier to move out of can be interpreted as a region of high curvature.

3.5 Human Perceptual Parameters for Haptic Interface Development

If we compare other human senses, such as hearing and sight, relative to touch, we realize that the understanding of the sense of touch is very limited as far as trying to recreate it in a hardware design context. Haptic devices try to recreate the stimulus that a real environment might exert on our skin and muscles. Our sensory receptors and cental nervous system are responsible for interpreting the sensation of touch, whether we are dealing with a real or simulated environment. Indeed, there are many reasons for this limitation. For instance, the experiential difficulty encountered when displaying controlled stimuli, e.g., force and torque, is mainly due to the fact that the haptic system is bidirectional, so it can simultaneously perceive and act upon the environment and can introduce mechanical instabilities under deterministic response [363].

Haptic interfaces used in realistic virtual reality simulations are designed with the understanding that they are operated by humans who have somatosensory and motor systems. In this context, many research prototypes and commercial products have been designed based on the known requirements for fooling the human haptic system, but in the strict sense, not all the interfaces are following the results of haptics research [99, 363]. In this subsection, we will deal with the design requirements of haptic interfaces, which can be characterized by two functions: measuring the position and contact forces of the user's hand and/or other parts of the body, and displaying contact forces and positions and/or their spatial and temporal distributions to the user.

Consequently, a good knowledge of human capabilities and limitations in both the sensory and the motor domains enables haptic device designers and manufactures to determine a reasonable level of haptic sensation accuracy in order to display more realistic touch stimuli. Therefore, a systematic exploration of the capabilities and limitations of the human sensory and motor systems constitutes a prerequisite to the appropriate design of any haptic interface.

The purpose of this section is to present human-centered system specification design guidelines for haptic devices being proposed and considered in research literature. These specifications are essentially haptic-relevant thresholds, or maximum values measured in humans, that are used as guidelines for haptic interface development. These include the fundamentals of temperature perception, which can add a complementary stimulus to the whole development of haptic devices.

3.5.1 Human Factors in the Design of Haptic Interfaces

Design criteria specifications of the hardware and software of haptic interfaces are mainly influenced by the fundamentals of biomechanical, sensorimotor, and cognitive abilities of the human haptic system. Knowing these abilities plays an important role in the understanding of how to build and control machines that display and meet haptic performance metrics. Perceptual characteristics, such as temperature perception, have been studied in order to be incorporated into haptic interfaces for object identification in virtual environments [33, 90, 196, 360].

The just noticeable difference (JND) is the smallest detectable difference between adjacent sensory stimulus levels. It has been intensively investigated and used as a quantity of the subject's sensitivity. More details about actual values and psychophysical evaluation methods can be found in [41]. The human factors that influence the quantitative performance of force-reflecting haptic interfaces have been studied by Tan et al. [363]. Those studies are related to ground-based controllers (for example, a stationary desktop controller) and body-mounted exoskeleton devices. The following perceptual elements are considered: (1) force sensing (2) pressure sensing (3) position sensing resolution (4) stiffness (5) human force control (6) temperature perception and (7) friction perception. In the following section, we will discuss these parameters.

3.5.2   Perception Thresholds

This section focuses on human perception thresholds for different physical parameters accessible to the haptic sensory system. These parameters has been chosen due to their relevance as important factors when designing haptic displays. For instance, for the user to perceive the forces displayed by the device as varying smoothly, the force display resolution of the device should match or exceed the human sensing resolution. Similarly, the vibrations generated by the haptic device should remain below a given threshold to maintain controllability of virtual objects and to avoid deterioration in the user's perception.

3.5.2.1   Force Sensing

The fundamental definition of forces based on Newtonian theories and mechanics have been fully replicated by laboratory equipment through experimental work. In order for the human user to perceive the force displayed by a device, its output force resolution should be equal to or exceed the human sensing resolution. In this context, a study presented by Jones in 1989 showed that body references sense forces in a range of 25–410 newtons (N); meanwhile, similar studies under different test conditions were carried out by reporting that force sensing resolution falls between 2.5 and 10 N [279]. However, some disturbances, such as vibration, can occur in force displays and deteriorate the force sensing quality. Vibration can be conceived as the level of mechanical oscillations that are related to the equilibrium reference. The equilibrium reference can be referenced when no net influences are acting on an object. In other words, the net force or net torque is equal to zero.

3.5.2.2   Pressure Perception

Pressure is a tactile sensation and proprioceptive or kinesthetic perception that refers to the awareness of one's body state when perceiving a continuous physical force exerted on or against it. The perceptual thresholds for touch depend on location, stimulus type, and timing states. Mechanoreceptor density varies across the skin, and tactile thresholds for single and two-point discriminations vary based on the skin area activated. It is suggested that actuator design must, therefore, consider the spatial resolution. For example, in order to apply two separate stimuli to the index finger, the tactile actuators should be at least 2.5 mm apart because humans cannot differentiate two stimuli which are closer than 2.5 mm to each other.

It is well known that the most sensitive body sites for single point localization are the nose and mouth, followed by the finger pad. Inversely, the most sensitive body site for two-point discrimination is the finger pad, followed by the nose or mouth. In addition, pressure results showed different levels of sensitivity dependent on body loci and gender, hence the JND values vary from 5 milligrams (mg) on a woman's face to 355 mg on a man's big toe [144].

On the other hand, kinesthetic devices, such as force-reflecting exoskeletons, are designed to display contact forces at the finger pad by attaching it to the user's forearm. In this context, Tan et al. [363] have applied an experimental work on the forearm in order to measure pressure as a function of contact area based on the JND concept. They found that the JND of pressure decreased at a ratio of 1:4 (from 15.6 to 3.7%) when the contact area increased by a factor of 16 (from 1.3 to 40.4 cm^2). These findings imply that humans are less sensitive to pressure changes when the contact area of the forearm is reduced. In addition, they suggest that the contact area of the exoskeleton's interaction points should be minimized, and the perimeter of the true grounding area should be maximized to enhance the overall illusion.

3.5.2.3 Position Sensing Resolution

When we talk about position sensing, we usually refer to single point interaction haptic devices; however, other types of haptic devices, such as force feedback gloves, can also be considered. Position sensing resolution can be defined as the smallest change on the end effector of a haptic device that can be detected in dpi or in millimeters. For example, the recognized Desktop PHANToM Device can provide a nominal position resolution >450 dpi (0.055 mm). In addition, researchers have found that this feature mainly depends upon the position resolution of the human operator. Tan et al. [363] have shown that joint angle resolution is directly related to the fingertip position, where the JND of proximal interphalangeal (PIP) and metacarpophalangeal (MCP) is related by 2.5°. In addition, measures of joint angle JNDs for the wrist, elbow, and shoulder joints were reported; they found that JND decreases from 2.0° at the wrist and elbow joints to 0.8° at the shoulder joint and that proximal joints present higher resolution in sensing joint angles than do distal joints.

3.5.2.4 Stiffness

In the design of haptic interfaces, the perception of stiffness requires a minimum threshold to emulate a rigid object in a virtual environment. One attempt by Tan et al. [363] reported results in an experimental study based on subjects pressing downward on a rectangular aluminum beam clamped at one end (see Fig. 3.11).

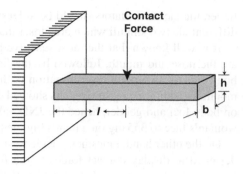

Fig. 3.11 Setup for an experimental study on stiffness (re-edited from [363])

Their results, based on the elastic beam theory formulation, have shown that at a given range of length l (say 31.0 ± 5.1 cm), the interval of K, defined as the "threshold" point, ranged from 153 to 415 $N\,cm^{-1}$. However, their experimental set up presented an issue related to the incremental step of distance l between the clamped end and the points located on the bar: the bending stiffness of the aluminum beam decreases at a function of $1/l3$.

Thus, they concluded that the minimum stiffness required to simulate a rigid surface is currently a mechanical design challenge. On the other hand, to get a better understanding of the forces involved in current haptic interface devices, the PHANToM Desktop can emulate stiffness around 1.86 $N\,mm^{-1}$ in the X-axis (horizontal), 2.35 $N\,mm^{-1}$ in the Y-axis (vertical), and 1.48 $N\,mm^{-1}$ in the Z-axis (depth).

3.5.2.5 Human Force Control

The maximum forces exerted by a device should meet or exceed the maximum forces humans can produce. It is then possible to establish the maximum controllable force a human can produce involving the arm, hand, and finger joints. The maximum sustained force exertion of the finger contact forces depends on the way objects are grasped, and the user's gender, age, and motor skills. The index, middle, and ring fingers exert about 7, 6, and 4.5 N, respectively, without fatigue or discomfort. Table 3.1 presents the results of a study based on measuring the maximum controllable force of subjects who were asked to close their eyes and exert a maximum force for at least 5 s. The test includes the PIP (proximal interphalangeal) and MCP (metacarpophalangeal) joints, the wrists, the elbow, and the shoulder. The maximum controllable forces fell in an interval from 16.5 N at the PIP joint to 102.3 N at the shoulder. In addition, the female subjects achieved a maximum controllable force range which is below that of the male subjects. Table 3.1 presents the highest scores of the force control reference values for a male subject's actions involving the arm, hand, and finger joints.

Table 3.1 Highest maximum controllable force (N) and standard deviation (%) based on the experiments by Tan et al. [363]

Subject	Joint tested				
	PIP	MCP	Shoulder	Elbow	Wrist
Male	50.9 N	45.1 N	102.3 N	98.4 N	64.3 N
	4.24%	4.47%	0.46%	2.47%	5.02%
Female	16.5 N	17.6 N	68.7 N	49.1 N	35.5 N
	3.99%	4.50%	3.67%	3.19%	3.12%

Table 3.2 Thermal display features based on the experiments by Jones and Berris [196]

Physical parameter	Thermal threshold
Maximum temperature range	20°C
Heating resolution	0.001°C
Cooling resolution	0.002°C
Number of elements in array	2–10
Temporal transient resolution – cooling	$20°C\,s^{-1}$
Temporal transient resolution – heating	$20°C\,s^{-1}$

3.5.2.6 Temperature Perception

Force and tactile feedback are the main sensory inputs presented to a human operator using a haptic display [196]. In addition, thermal feedback can be used to convey information about thermal conductivity of objects, which can help with object identification and creating a more realistic image of the object.

In this domain, the ability to perceive temperature depends on two different kinds of receptors found in the skin known as cold and warm receptors [196]. Cold receptors are more numerous than warm receptors by a ratio of up to 30:1, and they respond to decreases in temperature over a temperature range of 5–45°C. Warm receptors discharge due to an increase in skin temperature, reaching a maximum at around 45°C. In addition, when the skin temperature is maintained at 30–36°C, no thermal sensation is noted. The properties of the human thermal system, as a basis for specifying the desired features of a thermal display, are shown in Table 3.2.

3.5.2.7 Friction Perception

Resistance to motion in haptic interfaces has also been considered in the design and development of such devices. Thus, characterizing friction through quantifying human perception thresholds is an important factor in the quality metric for the design of haptic devices. Lawrence and his colleagues [217] described an approach for quantifying perceptual thresholds of friction that includes but is not limited to: the resistance defined under Coulomb friction; viscous and inertial forces; and mechanical imperfections. They have shown that observers perceived small differences around 0.2 μm at high frequencies. However, it was observed that "measuring the threshold of human perception for friction in a mechanical device is difficult because human perception of friction with stylus grips is mediated by at least three tactile psychophysical channels and perhaps a proprioceptive channel as well." It was also noted that all of these channels have sensitivities that are linked tightly with the frequency content of the stimulus.

3.6   Closing remarks

Haptic perception is typically characterized as the process of recognizing objects through touch. The sense of touch, in turn, is characterized by (a) the combination of somatosensory perceptions of patterns on the skin's surface (texture, edges, and curvature) and (b) proprioception, which refers to a person's sense of the relative locations of their body parts in space. In order to understand the complexity of this sense, many research fields have invested significant effort in the exploration of haptic characterization and representation.

A good starting point is the flow of sensory information through biological elements such as sensory receptors and sensory afferent neurons, which characterize the state of touch stimuli when interfacing with skin, muscle, and organs. These sensory signals propagate down to deeper neurons within the central nervous system and eventually to the brain, where the signals are processed. From here, we can further break down the study of the sensory system by identifying the different types of stimuli delivered through different sensory receptors; for example, mechanoreceptors provide tactile sensation and nociceptors react to pain stimuli. In addition, physiology research has also contributed to understanding how and where the brain processes the haptic sensory signals. Consequently, the study of haptic perception in this field has encouraged the new research domain of haptic technology, which emulates these touch stimuli in virtual and real environments.

Haptic perception in a real environment involves physical attributes, such as contact forces, when a human being explores a real surface. On the other hand, haptic perception in a virtual environment involves mechanical attributes that are generated through mechatronics systems or interfaces, known as haptic devices. Thus, when a user interacts with a virtual object, they receive haptic feedback in the form of device forces that characterize real objects. Future research into the sense of touch must recognize the complexity of reproducing this type of interaction.

Chapter 4
Machine Haptics

4.1 Introduction

The development of HAVE applications encompasses the development of both audiovisual devices and haptic devices to deliver a higher sense of immersion in a 3D space. 2D and 3D audio technologies have been introduced to create the illusion of sound sources placed anywhere in a three-dimensional space. By processing relative left and right speaker signals, apparent sound locations can be perceived at an arbitrary point in space. Visual information in HAVE applications can be characterized by the field of view (FOV), which represents the total visible angular deviation. The FOV needs to cover between 60 and 100° along the horizontal axis in order for the user to be immersed in the virtual environment. This is less than the capability of the human eye, which has an FOV range between 180 and 270°, depending on whether the eye is moving or not. The update rate for visual feedback is around 75 Hz, and the suggested resolution is on the magnitude of 1,960 × 1,280 pixels, even though it is possible to reach 8,000 × 8,000 pixels.

Most stereo vision systems are based on human binocular cues in order to render depth information. In the human binocular visual system, each eye captures its own image of the environment; the two images are then sent to the brain for processing. The brain combines them into a single picture by matching the similarities and compensates for the differences, which are usually small. The difference between the two captured images is the reason human beings can see a stereo image.

In HAVE applications, there are several ways to view stereoscopic images. In the first technique, a head-mounted display provides each eye with a separate image. The provided images can originate from a single video source or two different video sources. The FOV is made flexible through the use of head tracking. The same principle can be applied with a less bulky device known as shutter glasses or active glasses. In this case, users are asked to wear glasses and look at an appropriately configured system such as an LCD monitor. The shutters in the glasses, with the help of an infrared emitter, are than synchronized with the display system. Images are alternately displayed to each eye to provide different perspectives so that each eye

A. El Saddik et al., *Haptics Technologies*, Springer Series on Touch and Haptic Systems, DOI 10.1007/978-3-642-22658-8_4, © Springer-Verlag Berlin Heidelberg 2011

sees only the image intended for it (the glasses achieve this by alternately darkening over one eye at a time). Passive glasses, on the other hand, are color encoded or polarized glasses used to view a projected image. A stereoscopic system projects stereo video on the screen where the two images are tinted to different colors. Due to the color encoded glasses, each lens blocks the light color it is tinted to, so the image tinted to the opposite color makes it through to the eye. Finally, auto-stereoscopic displays are the new era in stereo vision. Users do not need to wear any glasses. In this type of system, lenses in front of or behind a display screen focus the image so that each eye sees a slightly different image. Auto-stereoscopic displays require careful calibration and adjustment, i.e. the user needs to be in front of the display and at given height and distance from the display.

Haptic device development is evolving in terms of shape, size, and mode of operation; such evolution is fueled by research and application requirements. The fundamental characteristic of a haptic device is the bidirectional principle of exchange of energy, whereby such devices both supply and dissipate energy to and from the system. The design and fabrication of haptic/kinesthetic feedback mechanisms is a virtually new field that will thrive with the advent of the wide spectrum of applications discussed in Chap. 2. This chapter first provides an overview of the traditional components that comprise a haptic interface, then explores the attributes that define the quality of haptic interfaces, and finally, offers descriptions of some existing haptic interfaces.

4.2 Haptic Interfaces

4.2.1 Robotics Perspective

The word "robot" is popularly associated with the stereotypes generated from science fiction films such as "Star Wars," "The Terminator," and "RoboCop." They are portrayed as fantastic, intelligent, and sometimes dangerous forms of artificial life. The development of robots ranges from very complex computer controlled devices, such as walking robots, to simple devices, such as toy robots, and take on a variety of forms and shapes. Generally, a robot is a programmable system connected to a mechanical structure with the main goal of performing designed manual tasks. The science involved in the design and development of robots is given the term 'robotics'. The manufacturing industry made great contributions to the development of robots from the 1960s to the 1980s. Their main concern was to find ways of increasing productivity while reducing costs of manufacturing products. However, the main issue in the development of robotic mechanisms was associated with force control, which includes the integration of task goals, trajectory generation, force and position feedback, and the modification of trajectories [393]. Whitney, in his survey, mentions that robot force control research began with remote manipulator and artificial arm control in the 1950s and 1960s.

With the advances in computer technology research and mechanical engineering, several applications have moved toward using computers instead of human operators. Such advances took advantage of evolving technologies, such as numerical control (NC) systems, computer-aided design (CAD), computer-aided manufacturing (CAM), computer numerical control (CNC), and computer integrated manufacturing (CIM), to create robotic work cells that perform the work on assembly lines without the use of human labor. In addition, the development of tele-operation and tele-robotic technologies played a key role in supporting physical action at a distance. This allowed physical actions, achieved through tele-operation, to change the state of remote systems, which could potentially be hazardous environments. For example, nuclear laboratories were concerned with the need for manipulating highly toxic materials in a safer manner. Thus, in the early 1950s, Goertz [133, 134] developed electric-servo manipulators with force reflection. These mechanical master–slave manipulators enabled a human operator to manipulate the master device with his/her hand and feel the contact forces experienced by the slave. These tele-manipulation systems are needed to protect human operators from radiation and other issues in a radioactive hot lab while they perform hazardous tasks. Later, the need for helping amputees to recover kinesthetic perception encouraged the researchers Rotchild and Mann to develop a force feedback powered artificial elbow for amputees [317]. In these robotic elbows, a joint motor was driven by signals from muscle electrodes, and a strain gauge in the joint enabled the amputees to exert muscle work in a similar fashion to how they had performed the same task with their natural arms.

In some cases, tele-operation systems can include force feedback so that the exchange of mechanical energy can be perceived directly by the human operator. This type of interaction is essential and can also be described as "kinesthetic" or "haptic" perception.

Thus, robots can be considered as haptic interfaces where they exchange mechanical energy with a human user by receiving commands and feeding back interaction forces. In other words, a human user exerts force onto haptic interfaces to move a virtual avatar of the user or a tele-operator; at the same time, the haptic interface feeds interaction forces back to the user when there is contact with something in the remote environment. Natural or manufactured objects can be seen as either inert, able to dissipate energy, or active. An active object is not only able to dissipate energy, but also able to supply energy to the system [152]. Based on such principles, Hayward et al. [152] classified haptic devices as either passive or active. They mention that passive devices are often designed to have programmable dissipation as a function of position or time. However, passive devices rely on nonholonomic constraints and are associated with the ability to modify, under computer control, the elastic behavior of an element so that it becomes harder or softer [152]. In order for the hand to perceive kinesthetic information of a manipulated object, such as position or movement, active devices exchange energy between a user and the machine as a function of the feedback control. Therefore, two categories can be conceived based on such a principle: impedance-type interfaces and admittance-type interfaces [152]. Impedance-type interfaces have their actuators acting as force

generators based on the measured position, whereas admittance-type interfaces have
their actuators generating positions based on the measured forces.

4.2.2 Haptic Interface System

The basic elements of a haptic system are the power supply, the computer controller,
and the physical device, also called the haptic display. In this book we focus only on
the technology that is used to measure the system response in terms of tactile and
kinesthetic cues and to actuate a haptic feeling.

A haptic interface system can be considered as a mechatronics framework that
includes one or more electromechanical transducers (sensors and actuators), as
shown in Fig. 4.1. Sensors and actuators convert energy in mechatronics systems,
and magnetic circuits seem to be the best medium for such conversions [284].
Sensors are employed to register and measure interactions between a contact surface
and the environment, whereas actuators provide mechanical motion in response to
an electrical stimulus. In other words, sensors measure mechanical signals, such as
positions, forces, or a combination of these and their time derivatives, to map the real
space into the virtual space; actuators apply mechanical stimuli at distinct areas of
the user's body (force signals) to approximate a realistic experience. In general, the
mechatronics framework includes analog-to-digital and digital-to-analog converters
(abbreviated as ADC and DAC, respectively) and a communication module.

The mechatronics framework communicates data between the transducers and
the computer system in a bidirectional manner. The ADC converts the data received
from the sensors to an equivalent digital value that is conveyed to the computer
system. The DAC does the opposite; it converts the digital commands into an analog
form (usually voltages) that is sent to the actuator(s). Usually, the exchange of
haptic data occurs at a very high rate, called the Servo Loop Rate, to provide a

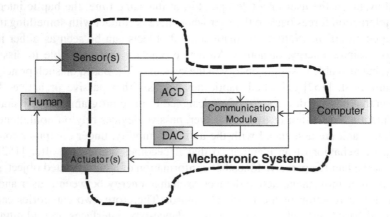

Fig. 4.1 Haptic interface system

more stable and realistic human–computer interaction. The communication module preprocesses the digitized data and implements the communication protocol that the haptic device uses to interface with the computer. For example, the communication module of the PHANToM Omni device implements a high-speed, serial input/output bus (the IEEE 1394 protocol) to connect the haptic device as a peripheral to the computer.

4.3 HAVE Sensors

Fundamentally, a sensor can be defined as a device that measures a physical quantity and converts it into a signal that can be read by an observer or an instrument. Sensors are usually sensitive to the measured physical property and do not influence it. The output signal of a sensor is a simple function (linear) or is linearly proportional (algorithmic) to the value of the measured physical property. Sensor readings can deviate from actual values in several ways. The most common deviations are:

- *Sensitivity error*: the measured value differs from the real value. This is a typical error that depends on the quality of a sensor. The lower the sensitivity error, the more precise and expensive the sensor is.
- *Offset or sensor bias*: this error occurs when the measured property is zero, but the output signal is not zero.
- *Dynamic error*: caused by a rapid change of the measured property over time, or in the case of digital sensors, through the sampling frequency.
- *Noise*: represents a random deviation of the signal over a given period of time.
- *Digitization error*: occurs with digital sensors when the input is converted to a digital signal. The output is an approximation of the real measured physical property.

Sensors can be used in a wide variety of applications to measure or detect temperatures, flow, vibration, force, radiation, etc. However, in the development of haptic interfaces, a sensor is mainly used to measure parameters of contact between the sensor and an object. The contact measurement is confined to a small, defined region and measures the fundamental attributes in the interface design: position, force, or pressure. A variety of sensors, such as piezoelectric sensors, force sensitive resistors, intrinsic or extrinsic fiber optic sensors, micro-machined sensors, and capacitive sensors are currently available with their own operating principles [303].

Knowing that the human finger resolution is about $40\,\mu$m, sensors are limited to being single-point sensors arranged in an array. Consequently, a tactile sensor comprises discrete sensor cells, called "texels," that are arranged in homogeneous matrices to detect an applied load profile. Sensor arrays are generally 10–15 rows of sensors with 1–$2\,mm^2$ resolution per sensor, which means every sensor monitors a region of around 1–$2\,mm^2$. The sensitivity of the touch sensor is generally considered satisfactory between 0.4 and 10 N, but this is dependent on

Table 4.1 Examples of array sensor implementations and their densities

Category	Sensing strategy	Implementation example	Density mm-2	Size
		[43]	0.18	8 × 8
	Capacitive devices	[348]	0.27	8 × 8
Electrical		[116]	0.07	8 × 8
	Piezoresistors	[356]	1.58	16 × 16
	Polysilicon piezoresistors	[362]	4.00	32 × 32
	Conductive plastic	[298]	1.00	6 × 3
		[160]	2.56	16 × 16
Mechanical	Conductive silicone rubber	[306]	0.69	16 × 16
		[345]	1.00	64 × 64
	Conductor rubber strain gauge	[321]	0.007	5 × 5
	Electrorheological	[261]	0.25	N/A
Magnetic	Magnetic dipole	[143]	0.25	7 × 7
Electromagnetic	Optical waveguide	[238]	0.08	10 × 10
	Ultrasonic	[179]	0.31	16 × 16

a number of variables determined by the sensor's basic physical characteristics. The measurement principles of tactile sensor cells are based on a variety of technologies; these including piezo devices, force sensitive resistors, intrinsic or extrinsic fiberoptic sensors, micro-machined devices, and capacitive devices.

Table 4.1 lists some existing implementations suitable for static array sensing. It is worth mentioning here that when choosing a particular technology, one must keep the intended HAVE application in mind. For instance, in manipulation scenarios where a round finger is intended to roll around an explored object, some technologies, such as conductive plastic or polysilicon piezoresistors, are not suitable for the geometry.

4.3.1 Electromechanical Sensors

Early sensing technologies were purely mechanics-based until the introduction of electromechanical technologies. These new technologies enabled the conversion of mechanical information into equivalent electrical signals. For example, electro-mechanical sensors are commonly used in automotive industry applications, as they are characterized by high reliability and sensitivity. They are mainly used to build tactile sensing arrays; however, they suffer from being fragile and vulnerable to overpressure due to mechanical compliances, as well as being too expensive and bulky for use in wearable applications. Furthermore, electro-mechanical sensors are vulnerable to electromagnetic interference (EMI) and corrosion.

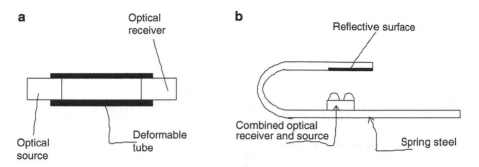

Fig. 4.2 (**a**) Force sensitivity touch sensor, (**b**) reflective touch sensor

4.3.2 Optical Sensors

The main benefits of optical sensors are that they are immune to external electro-magnetic interference, do not induce physical damage, and are small and light. The operating principles of optics-based sensors are divided into two classes: intrinsic and extrinsic. The following is a brief explanation of both classes.

Intrinsic fiberoptic sensors imply that the sensing takes place within the fiber itself. Intrinsic-based sensors utilize force-dependent absorption or reflection of light beams [303]. As shown in Fig. 4.2a, moving an obstruction into the light path causes a modulation of the intensity of light. Notice that the force sensitivity is determined by a spring or elastomer. Intrinsic types are related to applications that measure rotation, acceleration, strain, acoustic pressure, and vibration measurement. They are more sensitive, but they are more expensive and tougher to multiplex. On the other hand, extrinsic fiberoptic sensors are distinguished by the fact that sensing takes place in a region outside the fiber. The physical stimulus interacts with the light external to the primary light path. For instance, in the reflective touch sensor shown in Fig. 4.2b, the intensity of the received light is a function of the distance between the reflector and the plane of the source and represents the applied force. The U-shaped spring can be manufactured from spring steel, leading to a compact overall design. Based on the intensity measurement, extrinsic fiberoptic sensors are most widely used due to their simple structure and information processing. In general, extrinsic sensors are less expensive, easier to use, and can be assembled to support arrays of sensors; nonetheless, they are less sensitive.

4.3.3 Capacitive Sensors

Electrical sensors produce a change in electrical or magnetic signals based on an environmental input. The most well-known electrical sensors are radar systems, metal detectors, and electrical meters such as ohmmeters. Capacitive sensors utilize

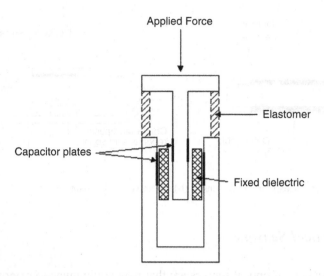

Fig. 4.3 Capacitive tactile sensor

the change of capacitance between two electrodes covering a deformable dielectric [295]. This can be achieved by either changing the distance between the electrodes or changing the effective surface area of the capacitor (as shown in Fig. 4.3). The two electrodes are separated by a dielectric medium, which is used as an elastomer to give the sensor its force-capacitance characteristics. One of the limitations of capacitive sensors is that there is an effective limit on the resolution of the capacitive array. Another issue is the need for complex signal conditioning (filtering and amplification) to detect very slight changes in the capacitance. Filtering is needed to improve the signal-to-noise ratio (SNR) of the output signal. Amplification is then used to increase the signal strength for data acquisition, transmission, and processing [350].

4.3.4 Resistive Sensors

The use of compliant materials with defined force-resistance characteristics has received considerable attention in touch and tactile sensor research. The basic operation of this type of sensor is based on the measurement of the resistance of a conductive elastomer between two points. The majority of these sensors use an elastomer that consists of a carbon-doped rubber. Generally, due to their simple structure, resistive tactile sensors are very robust when withstanding overpressure, shock, and vibration. As shown in Fig. 4.4, upon the application of external forces, the deformation of the elastomer alters the elastomer material density, thereby changing the resistance of the elastomer. Consequently, after a given period of time, the elastomer will become permanently deformed and fatigued, leading to

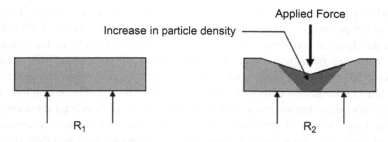

Fig. 4.4 Resistive-based tactile sensor

permanent deformation of the sensor. This impacts the sensor's long-term stability and means it requires replacement after an extended period of use.

4.3.5 Force Sensors

Another possibility is the use of a Force Sensing Resistor (FSR). It is a piezoresistivity-conductive polymer that changes resistance in a predictable manner following the application of force to its surface. It is normally supplied as a polymer sheet with the sensing film applied to it by screen-printing. The sensing film consists of both electrically conducting and nonconducting particles suspended in a matrix. The particle sizes are of the order of fractions of microns and are formulated to reduce the temperature dependence, to improve mechanical properties, and to increase surface durability. Applying a force to the surface of the sensing film causes particles to touch the conducting electrodes, which changes the resistance of the film.

4.3.6 Strain Gauge Sensors

A strain gauge detects the change in length of the material attached to it when external forces are applied. In HAVE applications, the strain gauge can be used as a load cell where the stress is measured directly at the point of contact, or positioned within the structure of the end-effectors to measure the applied force.

4.3.7 Magnetic Sensors

Usually, magnetic sensors are used to sense in one direction. They are based on the movement of a small magnet by a given applied force, which causes the flux density at the point of measurement to change. The flux measurement can be made by either a Hall effect or a magnetoresistive device [264]. The Hall effect is the production of a voltage difference (also known as the Hall voltage) across an electrical conductor according to an electric current in the conductor and a magnetic field perpendicular to the current. A magnetoresistive material is a material whose magnetic characteristics are modified when the material is subjected to changes in externally applied physical forces. The magnetoresistive (or magnetoelastic) sensor has a number of advantages, such as high sensitivity and dynamic range, a lack of measurable mechanical hysteresis, a linear response, and physical robustness. Examples of array sensor implementations and their densities are shown in Table 4.1.

4.4 HAVE Actuators

An actuator is a mechanical device used for moving or controlling a mechanism or system. In haptic device design, an actuator is a force and/or position source that exerts forces on the human body/skin to simulate a desired sensation. Abstractly, actuators are used to change the impedance state of the device between virtual free space and virtual constraints. The most important factors in actuator design are: the speed of operation (response time), safety, mechanical transparency, workspace, number of degrees of freedom, maximum applicable forces and stiffness range, compactness, and control bandwidth. In HAVE applications, we distinguish between many different types of actuators, including electrical, pneumatic, hydraulic, piezo-electric, and memory-alloy among others. In this section, we will discuss the three most important actuators for HAVE applications.

Electrical actuators: include motor-based actuators with many different types of motors, such as direct current (DC), brushed, permanent magnet (PM), stepper motors, and rotary, linear, and latching solenoid actuators. They do not require significant amounts of space to operate and are easy to install, as there is no complex wiring, and no pump rooms are needed. They produce only negligible levels of electromagnetic noise oscillation that can interfere with other communication devices located in the peripherals, and are usually easy to control due to their solid state electronics. The disadvantages of electrical actuators are the small torques they generate (compared to their size and weight), their low bandwidth, and rigidity (they do not bend and thus cannot be embedded in wearable devices).

Pneumatic actuators: utilize compressed air pressure to transfer energy from the power source to the haptic interface; they are technically simple and lightweight. They provide higher power-to-weight ratios than electrical actuators. Since they use air, the device may be used in clean environments. Their disadvantages include low bandwidth and stiffness due to the compression of air. Also, the use of dry air in pneumatic actuators means lubrication is an issue because static friction is not handled well.

Hydraulic actuators: based on a fluid, which is in most cases oil. Due to the nature of fluids as self-lubricating materials, hydraulic actuators are considered to be high bandwidth devices that do not suffer from the friction problem found in pneumatic devices. They are powerful enough to support heavy payloads and haptic force interactions. The disadvantages are their bulkiness and weight. Another disadvantage is their need for more maintenance since the oil must be filtered and cleaned on a regular basis.

Many different design approaches have been investigated in order to optimize the performance parameters of haptic interface actuators. Examples of existing approaches include magnetic levitation devices, nonholonomic displays, cable-driver linkage and tensed string systems, parallel mechanisms, and ungrounded and exoskeleton-type interfaces. The most widely used are motor-based serial devices. In the following section, we provide a brief summary of these approaches and their relative benefits and limitations.

4.4.1   Magnetic Levitation Devices

Magnetic levitation devices use the Lorentz force principle to suspend an object with the support of magnetic fields. An electric current in a magnetic field generates the method of actuation. The levitated part of these devices is referred to as the floter, while the stationary base containing the permanent magnetic assemblies is the stator. The use of magnetic levitation for haptic interaction has been demonstrated in the IBM Magic wrist [172], the UBC wrist [396], and many others [36, 326]. The approach is simple, compact, and offers the potential to achieve relatively high bandwidths. On the other hand, the major shortcoming is that it has limited workspace.

Butterfly Haptics adopted the principles of magnetic levitation and developed the Butterfly Haptics Maglev 200 device [173]. The device consists of a handle that is rigidly connected to a hemispherical flotor. The flotor is freely levitated in magnetic fields generated by the stator. The device is a six DOF force feedback device (position and orientation feedback) with a peak force feedback of 40 N. The device workspace is a 24 mm diameter sphere (for translational workspace) with a ±8° rotational range. A snapshot of the device is shown in Fig. 4.5.

4.4.2   Nonholonomic Devices

The application of traditional robot manipulator control techniques for haptic displays has unearthed some challenging issues [80]. First, developing a perceptually smooth force-controlled motion is not a trivial task and requires relatively high servo rates. Second, stability and safety are important issues when a device must

Fig. 4.5 The Butterfly
Haptics Maglev 200 device

physically interact with a user. As an alternative, passive or nonholonomic displays offer the potential to alleviate such problems of performance, stability, and safety. Nonholonomic displays enable virtual constraints to be implemented in a manner that is completely passive and intrinsically secure and safe [82]. The idea is to begin with a device having zero or one degree of freedom (DOF) and to use feedback control to increase the apparent degrees of freedom as necessary. This is made possible thanks to nonholonomic joints, which have fewer degrees of freedom than generalized coordinates. Examples of designs that utilize nonholonomic displays can be found in [82, 286].

4.4.3 Magnetic Sensors

Tensioned cable systems were initially developed at Jet Propulsion Laboratory (JPL), the Tokyo Institute of Technology [331], and the University of Texas in Austin [230]. In cable-driven systems, the user grasps a handle that is controlled and supported from all directions by several actuated cables or springs. The combined tension exerted in the cables produces a net force and/or torque on the user's hand. The workspace can be made very large while the actuated inertia remains relatively small. Cable and linkage based devices have the advantage of being lighter and exhibiting less susceptibility to friction and backlash. Furthermore, the kinematics and control of the device are well understood. On the other hand, due to the limitation in the mechanical properties of the linkage, such actuators have a very limited bandwidth. A commercially available example of a cable-driven haptic interface is the PHANToM device [246], which is probably the most widely employed haptic device currently in the field. Other examples of linkage-based devices are described in [188] and [78].

4.4.4 Parallel Mechanisms

In applications where workspace size is not a major concern, parallel mechanisms [178] provide an attractive alternative to traditional serial link mechanisms. A parallel mechanism is an appropriate candidate for haptic devices since it is well known for its high stiffness and low inertia, which allows large bandwidth force transmission. In a parallel mechanism, actuators are kept at the base of the device. Therefore, they are mostly grounded, which leads to lower device inertia and greater strength and rigidity. The main drawbacks of such devices are the workspace requirements, the high complexity of dynamic models, and the forward kinematics, especially when a high DOF is required. Examples of haptic devices that use parallel mechanisms are described in [374] and [379].

4.5 Performance Specifications

The existence of numerous operating principles for haptic devices implies that HAVE application designers and developers must find the best device to use for their specific application and context. Therefore, developers should investigate the various specifications of these devices. The purpose of this section is to explain some semantics, capture the meanings of different performance specifications, and present the psychophysical evidence that helps developers decide what is necessary in a haptic interface. In general, there are no typical values for these attributes, as it always depends on the specific application.

The performance specifications are divided into three categories: physical, spatial, and temporal. As one would expect, physical specifications are those associated with the mechanical attributes of the device, such as the exertable forces, inertia, stiffness, friction, backdrivability, etc. Spatial specifications are those that define the geometric features and behavior of the haptic interface. Major attributes in spatial specifications include the degrees of freedom, position resolution, precision, workspace, location, etc. Finally, the temporal specifications refer to the measurement of the haptic device's performance in real time. Examples of temporal attributes are the device latency, the haptic refresh rate, and the maximum acceleration.

4.5.1 Physical Attributes

1. *Inertia.* This attribute depends on the mass of the haptic device. The goal is to improve the transparency of the device by decreasing the inertia felt by the user in unconstrained free movement. In other words, the haptic device must enable communication between the real world and the virtual world without introducing extra forces resulting from the weight of the device. Usually, a control algorithm

is in place to try to compensate for the device inertia by generating extra forces. Another way is to install counter masses in the device to offset the mass of linkages. Translational inertia is measured in grams and rotational inertia is expressed as mass times a unit area.

2. *Backdrivability*. This refers to the ability to move the end-effector of the device within the workspace without opposition/resistance. Ideally, the device should generate no forces on the user's hand during free movement since there is no interaction with objects in the virtual environment. The device's backdrivability is usually the result of friction in the gears, the motors and their cable transmissions, and inertia represented as backdrive friction and expressed in newtons. Since these frictions can be different for varying degrees of freedom, the backdrivability might have different values for each DOF, especially in translation and rotation motions.

3. *Friction/damping*. Kinetic friction comes in two forms: Coulomb friction, and viscous or damping friction. Both forms are considered as forces of resistance that oppose motion. Coulomb friction is independent of velocity and is measured in newtons, whereas viscous friction is proportional to the velocity and is expressed as a coefficient in $N\,s\,m^{-1}$ or $kg\,s^{-1}$, so that when multiplied by the velocity, it yields a force in newtons. Improving the design of the hardware or compensating for unwanted forces by proposing control algorithms can reduce friction effects.

4. *Exertable force attributes*. This is essentially a bundle of attributes that characterize the ability and flexibility of the device to generate force feedback. These attributes include, but are not limited to, maximum exertable force, continuous force, minimum displayed force, and dynamic force range. The maximum exertable force illustrates the maximum force that the actuators of a haptic device can generate over a very small time interval (several milliseconds). The continuous force attribute describes the force that the hand controller can exert for an extended period. The minimum displayed force represents the force sensitivity of the haptic interface and depends on the device's ability to display very slight forces through low friction and precise motor control. Finally, the dynamic force range can be defined as the ratio of the maximum displayable force to the minimum displayable force. The larger the range, the better the device because it will have a greater ability to generate a wide variety of forces and torques in the virtual environment.

5. *Stiffness*. Stiffness is the ability of a device to mimic a solid virtual wall or object. This attribute is of particular importance to the perception of rigidity. In other words, stiffness is the required parameter to convey to users that an object is rigid. It is interesting to know that the required stiffness to perceive rigidity is higher when vision is obscured. According to a report by [363], stiffness needs to be $25\,N\,mm^{-1}$ to feel stiff to a user when vision is obscured, whereas lower values are sufficient when the optical system is not obstructed.

6. *Size/weight*. The size and weight of a haptic device has a direct impact on the comfort level of the user. Furthermore, the trend of integrating haptics into mobile devices, or the development of mobile haptic devices, is limited by the

weight and size of the device. In some cases, a larger size is unavoidable since large areas of the human body must experience stimuli to provide the intended perception.

4.5.2 Spatial Attributes

1. *Workspace.* The area or volume in real world space that the end-effector of a haptic device can reach is referred to as the workspace. The haptic workspace is classified as either translational or rotational. Translational workspace is the volume/area traversed in Cartesian coordinates, with a reference point located at the center of the workspace. This reference point is mapped to the center point of the reference system in the virtual environment. On the other hand, rotational workspace is measured as the angle range in pitch, yaw, and roll. The workspace should be chosen according to the intended range of motion.
2. *Position resolution.* Defined as the smallest amount of movement over which the position sensors can detect a change in the position of the end-effector point of the device. Essentially, the position resolution depends upon the human position resolution. A commonly accepted measure of human sensory resolution is the just noticeable difference (JND). Usually, the positional resolution requirement depends heavily on the application. For instance, gaming and entertainment applications can tolerate lower position resolution in exchange for larger forces and higher force ranges. However, in applications such as surgery, a higher accuracy of the stylus movement is needed.
3. *Degree of freedom.* The term degree of freedom (DOF) is used to describe the haptic interface motion, sensing, and actuation capabilities. They do not always correspond to the number of joints. For instance, the sensing DOF refers to the number of independent position variables needed in order to locate all parts of a mechanism. As per the actuation of the DOF, it has been used to refer to the number of independent directions along which the device is able to display forces and/or torques. Available devices range from those capable of producing only nondirectional forces such as vibrations to six DOF devices that can activate forces along and around all three spatial axes.
4. *Precision and Repeatability.* Precision refers to how accurately the position sensor can refer to its position. This attribute is different from the position resolution since it does not involve any motion. Repeatability represents how accurately the haptic device can sense the identical physical position as being the same virtual position. This parameter is important in scenarios such as haptic playback.
5. *Grounding location.* The grounding location is the base reference that the device is attached to. It can be ground based, body based, or un based. For instance, ground-based haptic interfaces have a relatively stationary aspect and incorporate a heavy base to lend stability against the application of grounding forces. Examples of ground-based haptic displays are the PHANToM and HapticMaster.

Body-based devices, such as the Cybergrasp force feedback glove, are typically designed to be held or worn. Un-based interfaces, such as the AirGlove, neither use the ground nor the user's body for force reaction, but instead use other means, such as angular momentum, for force reflection.

4.5.3   Temporal Attributes

1. *Device latency*. The device latency, or delay, is the time measured from the instant of sending a command to the device to the instant of receiving a response from the device. This might include geometrical computation delays that may originate in software processing time.
2. *Bandwidth*. The bandwidth of a haptic device is defined as the range of frequencies over which the hand-controller provides force feedback. Theoretically, the desired bandwidth depends on the human perception system, but practically, it depends on the operation performed. Generally, precise and small movements require a higher frequency feedback than larger, more powerful movements. This is why tactile forces require a higher frequency bandwidth (100–10,000 Hz), whereas kinesthetic sensing requires a smaller bandwidth (20–50 Hz). The response band required by the hand and fingers is much lower (around 5 Hz).
3. *Haptic refresh rate*. This rate is the speed at which the feedback loop can be completed, and it is usually expressed in hertz. This includes continuous execution of the haptic rendering algorithms from sampling of the position sensors to the application of corresponding reaction forces on the operator through the haptic device. Although there are no firm rules for the haptic refresh rate, as noted in Chap. 3, 1 kHz is a common value. Notice that increasing this rate increases the realism of human–computer interactions, but only at the expense of increased computational power or reduced scene complexity.
4. *Maximum acceleration*. This attribute reflects the ability of a haptic device to simulate stiffness of virtual objects such as walls. It can be measured using an accelerometer attached to the hand controller.
5. *Haptics update rate/system latency*. The system latency is the total time delay of the haptic virtual reality system. This includes sensing the position of the haptic device, computing the force feedback in the simulation, sending the force to the device, and reading the next position. The haptics update rate depends on the quality of haptic rendering, the speed of the computer, and the type of haptic device. The inverse of this delay, or the haptic update rate, is quoted more often.

4.6   State-of-the-Art Haptic Interfaces

Classifying haptic interfaces varies by perspective. One important (and probably the most common) distinction among haptic interfaces is whether they generate a tactile or kinesthetic stimulation. Generally, kinesthetic haptic interfaces can be classified by their grounding locations (reference based), or by the degrees of freedom with which they can move. In this section, we adopt a hybrid classification that combines the force type (tactile/kinesthetic) and the actuation degrees of freedom. First, we classify haptic interfaces as either tactile or kinesthetic and then classify tactile and kinesthetic interfaces based on their DOF characteristic [106]. We present a cross-section of existing haptic device designs, selected to illustrate the diversity of designs among both the industrial and the academic communities. Please note that a complete survey would be much larger. We briefly describe specific technical requirements and specifications and comment on prominent features of these designs.

4.6.1 Tactile Interfaces: Commercial Development

Tactile stimulation refers to the sense of natural physical contact with the ambient environment. Tactile interfaces are devices capable of reproducing:

- Tactile sensations, such as pressure, texture, puncture, thermal properties, softness, and wetness
- Friction-induced phenomena, such as slippage, adhesion, and micro failures
- Local features of objects, such as shape, edges, embossing and recessed features

Currently, there are various commercial products and research prototypes that provide a diverse range of tactile sensations, some of which are presented in Table 4.2.

Several kinds of receptors have been found to mediate tactile sensations in the skin or in the subcutaneous tissues. Many proposed mechanisms use mechanical needles that are activated by electromagnetic technologies (such as solenoids and voice coils), piezoelectric crystals, shape memory alloys (SMA), pneumatic systems, and heat pump systems. Other technologies are based on electrorheological fluids, which change viscosity and rigidity upon the application of an electric field. Medical-specific technologies, such as electro-tactile and neuromuscular stimulators, are still under development.

A survey of tactile interface devices developed so far can be found in [172], but it is beyond the scope of this book. Currently, few tactile interface devices are commercially available, but they include the Touchmaster by EXOS Inc., the Tactool system by Xtensory, and the Teletact Glove by Intelligent Systems Solutions. Examples of tactile displays research prototypes are: HAPTAC from the Armstrong laboratory, linear and planar graspers developed by the Touch lab at MIT, a temperature display at Hokkaido University, a prototype tactile shape display at Harvard University, a programmable tactile array by the TiNi Alloy Company, and a tactile feedback glove at the University of Salford.

Table 4.2 Commercial and development tactile–haptic interfaces

	Product	Description	Sensation	Vendor
	CyberTouch	Vibro-tactile stimulators: six (one on each finger, one on the palm)	Pulses or sustained vibration	Immersion Corporation
	Touch Master	4 Vibrotactile stimulators (each finger)	Vibration	EXOS, Inc. (Microsoft)
	Tactile Mouse	Vibro tactile	Vibrations	Logitech
Commercial	Tactool System	2 Fingers	Impulsive vibration	Xtensory, Inc.
Interfaces	Displaced Temperature System	Via thimble	Temperature change	CM Research, Inc.
	HAPTAC	Tactile feedback	Electric pulses shape memory alloy (SMA)	Armstrong Laboratory
	Prototype Tactile Shape Display	Two-fingered hand with 2 DOFs in each finger	Electric pulses (SMA)	Harvard University USA
Research and	Temperature Display	Fingertip bed	Temperature feedback	Hokkaido University Japan
Development	Electrorheological fluids for tactile displays	Colloidal dispersion of malleable oil and dielectric solid particulate	Oil malleability	Hull University, UK
	Tactile display with flexible endoscopic forceps	Distal shaft of the forceps	Contact pressure sensations	Research Center at Karlsruhe, Germany
	Tactile Display	Thumb, index finger, middle finger, and palm simultaneously	Tactile stimulus	Sandia National Laboratories

4.6.1.1   The Tactile Mouse

Tactile feedback adds another cue in response to a user's action; it is one that can be felt even if the user is looking away from the computer screen. A tactile mouse helps a user to haptically distinguish graphic elements such as menu options, icons, or virtual objects by making them feel different when overlapped by the mouse cursor. Powered by these advantages, a number of tactile mice have been introduced to the market. For instance, the iFeel Mouse [3], illustrated in Fig. 4.6a, has an outside appearance, weight (132 g), and price (<$40) similar to those of a standard computer mouse. The only difference is that an electric actuator is attached to the mouse's body and can vibrate the mouse's outer shell. As shown in Fig. 4.6b, the actuator shaft translates up and down in response to a magnetic field produced by its stationary element. The actuator is oriented perpendicularly to the mouse base so that the vibrations occur in the vertical direction.

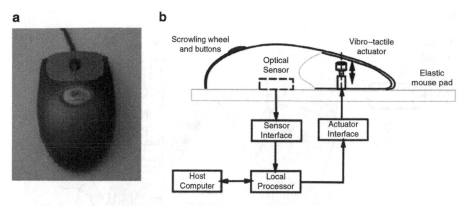

Fig. 4.6 The iFeel tactile feedback mouse: (**a**) outside appearance, (**b**) tactile feedback system

Furthermore, the iFeel mouse uses optical position measurements rather than the traditional mechanical ball. The reason is that vibrations produced by the actuator might interfere with the ball-roller assembly used to measure the XY-coordinates. The operation of a haptic mouse is simple: when the software detects contact between the screen arrow controlled by the mouse and the haptically enabled graphic items, it sends a haptic command indicating the onset and type of tactile feedback to the mouse processor. The processor converts these commands into vibrations with the desired amplitude and/or frequency and drives the actuator via its interface.

4.6.1.2 CyberTouch Glove

Released in 1995 by Virtual Technologies Inc., the CyberTouch is a haptic interface that provides vibrotactile feedback to the user. As shown in Fig. 4.7a, six tactile actuators (one on the back of each finger and one in the palm) are used to provide impulses and vibrations. These actuators can be used individually or in combination to produce synchronized tactile patterns. Each actuator consists of a plastic capsule housing a DC electrical motor. The motor shaft has an off-centered mass, which produces vibrations when rotated. Hence, by changing the speed of rotation, the vibration frequency can range from 0 to 125 Hz. Each actuator applies a small force of 1.2 N.

The functional building block diagram of the CyberTouch glove is shown in Fig. 4.7b. When the fingers and/or the palm of the avatar of the human hand interact with objects populating the virtual environment, the computer sends commands through the serial RS232 interface to activate the vibrotactile actuators. These signals are received by the driver unit, which sends the corresponding currents using the D/A converter and operational amplifiers to drive the motors. Due to its ability to provide feedback to individual fingers, the CyberTouch glove is most suitable for dexterous manipulation tasks where contact is at the fingertips.

a

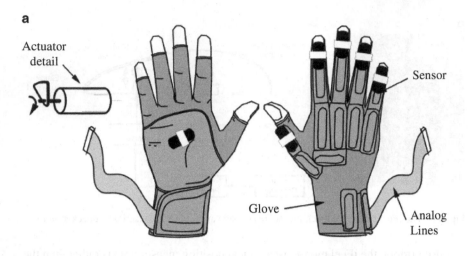

b

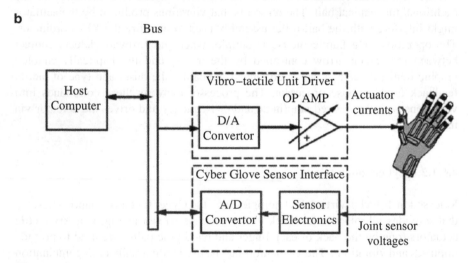

Fig. 4.7 The CyberTouch glove (**a**) outside appearance, (**b**) system diagram

4.6.1.3   The Displaced Temperature Sensing System

The Displaced Temperature Sensing System(DTSS) is a commercial haptic interface that provides temperature feedback for virtual environment simulations. The interface allows users to sense thermal characteristics such as surface temperature, thermal conductivity, and diffusivity, which can help in identifying an object's material properties. For instance, materials that have high conductivity (such as aluminum) will feel cold when touched, while those with low conductivity (such as wood) will feel warmer. This is due to the heat flow between the finger and the touched object.

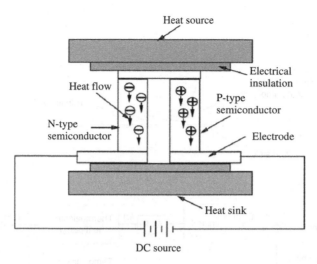

Fig. 4.8 Thermoelectric heat pump element

To increase the user's freedom of motion, and to support fast temperature changes, the actuators in DTSSs are thermoelectric heat pumps that function according to the Peltier principle. This principle stipulates that applying a DC current to a connected dissimilar material creates a temperature differential. Peltier pumps consist of solid-state N and P-doped semiconductors sandwiched between ceramic electrical insulators that act as thermal conductors and mechanical support. This is illustrated in Fig. 4.8. One of the plates is called the heat source while the other acts as a heat sink. When a DC current is applied to the heat pump, the P and N charges move to the heat source plate, which results in a rise in temperature in the heat sink plate.

One of the common DTSS models is the X/10, developed by CM Research. This DTSS model consists of a controller, eight thermodes, and connecting cabling. The controller can be programmed appropriately for input or output channels. It can be operated directly from the controller unit or through a computer via an RS232 serial interface. The temperature differential between the target and actual fingertip temperature is fed to the Proportional-Integrative-Derivative (PID) controller, as shown in Fig. 4.9. The output of the PID controller is sent to current amplifiers that drive the thermoelectric heat pump, and the control loop is closed.

4.6.1.4   Other Tactile Commercial Interfaces

The TouchMaster, introduced by EXOS Inc., is a tactile interface that allows the simulation of each of the four fingers and the thumb using electromagnetic voice coil actuators. These actuators are mounted on a cable assembly and attached to the fingertips using Velcro bands (commercial brand name for nylon fabric used as

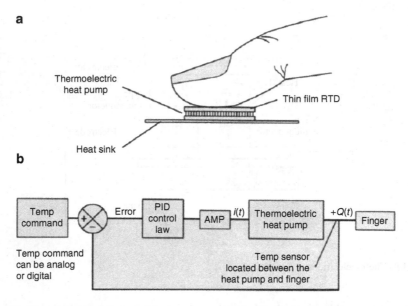

Fig. 4.9 Thermal control diagram (**a**) system components, (**b**) system diagram

a fastening) and are driven by a signalcondition box that interfaces the device to
the PC or any other standard digital I/O bus. The standard configuration provides
a vibration frequency of about 210–240 Hz at a constant amplitude. However, the
device is extendable in that optional variable frequency and amplitude electronics
are now available.

The Tactool system, developed by Xtensory Inc., is made of at least one tactor
connected by a cable to a power supply. The spacing between the pins is 3 mm.
Each actuator delivers a 0.3 N force and operates in vibration/oscillation mode. The
primary interface is serial EIA 232, but parallel, analog, and MIDI interfaces are
also available.

4.6.2   Tactile Interface Research Prototypes

In addition to commercially available tactile interfaces, there exist active groups
performing research and development of tactile interface prototypes. We will now
discuss some of these research prototypes.

At Sandia National Laboratories, the research group of Andaleon is working on
tactile devices for VE applications that interact with fingertips. The tactile interface
consists of a 2×3 pin-matrix of electromagnetic actuators mounted on a pad frame
and fixed on the finger of the user. Each actuator operates in the range of 8–100 Hz,
is capable of 762 μm indentation, and has a maximum pressure of $1.2\,\mathrm{N\,cm^{-2}}$.

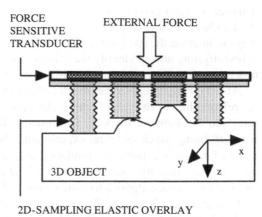

Fig. 4.10 Tactile sensor interface

Furthermore, each actuator can be controlled individually in terms of amplitude, frequency, and phase. The supporting software allows tactile displays to be used on the thumb, index finger, and palm simultaneously.

At the University of Ottawa, Petriu and his colleagues introduced a tactile sensor with a high sampling resolution (1.58 mm pitch) on a slave robot for active perception of stationary polygonal objects [402]. The experimental tactile interface consists of a 16 × 16 matrix of force-sensing resistor elements and an elastic overlay with protruding tabs that provides the spatial sampling, as shown in Fig. 4.10. Furthermore, a model-based method for blind tactile recognition of 3D objects is proposed in [287]. The geometric symbols representing terms of a pseudorandom array (PRA) are embossed on the object's surface. The method was tested on two 3D polygonal objects: a cube and a parallelepiped. In [285], they reported on the development of an intelligent multimodal sensor system to enhance the haptic control of robotic manipulations of small 3D objects. The sensor system is mounted on the end-effector of a manipulator arm with a relatively high resolution of 1/16 in (detects position change within 1/16 in.) and refines laser range 3D maps of fine interaction scenarios.

Researchers, led by professor Hassen at Armstrong Laboratory, have studied the perception characteristics on a tactile surface of a 5 × 6 array of actuators separated by 3 mm in each direction [149]. The actuators are based on shape memory alloy (SMA) wires that push/pull the tactile elements. The tactile array has been adapted for use in a device called the HAPtic-TACtile (HAPTAC). The HAPTAC device has been used in the TacGraph system to present data plots to blind persons.

At Harvard University, researchers working with Robert Howe have developed tactile interface prototypes that deliver shape and vibration feedback. They have conducted a series of experimental studies using these displays. The tactile interface uses pins driven by SMA wire actuators. The pin diameter is about 1.7 mm, the distance between two pins is 2.1 mm, the force delivered by each pin is 1.2 N, and the bandwidth of the display is 6–7 Hz [212]. Current research includes looking at ways to increase the bandwidth of the display to around 25 Hz. In addition, the team is investigating how to identify the dynamic range requirements needed for different tasks and to develop a detailed specification for system performance.

At MIT, researchers in the Touch Laboratory are looking at human haptics and its relationship to machine haptics. As part of a project that focuses on tactile interface development for VEs, the Touch Lab has developed two major devices for performing psychophysical experiments: the Linear and Planar Graspers. The Linear Grasper is capable of simulating mechanical properties of objects, such as compliance, viscosity, and mass, during haptic interactions, whereas virtual wall and corner software algorithms were developed for the Planar Grasper [236].

Researchers at Karlsruhe Research Center in Germany are developing a tactile feedback system for use with flexible endoscopic forceps [136]. This haptic interface is composed of 72 needle actuators placed in a 3×24 matrix. The individual needles are electromagnetically triggered by opto-decoupled printout boards and vibrate at a maximum frequency of 600 Hz to simulate contact pressure sensations.

Professor Ino and his team at the University of Hokkaido in Japan carried out research on tactile interfaces for presenting shearing and pressure forces and temperature feedback based on pneumatic actuators [187]. The pressure on each actuator (cylinder) is computer controlled by means of an electro-pneumatic regulator. This controls both the pressure and shearing sensations generated by means of a lateral-moving stage (displacement). The amplitude of the displacement is 3 mm in both the X and Y axes. Moreover, another interface has been developed to provide thermal feedback [187]. A thermocouple measures the temperature of the display surface and enables the display to act as either a cooler or a heater using a Peltier module.

Researchers at Hull University have investigated the use of electrorheological fluids for tactile displays. The fluids are primarily a colloidal dispersion consisting of insulative base oil with a slightly conductive dielectric solid particulate. Upon applying an electric field, these fluids have the ability to change from a liquid to a pseudo-solid state almost instantly, and their malleability is dependent on the strength of the applied field. Researchers have proposed and tested a tactile interface, with a 5×5 actuator matrix, for HAVE applications [367].

In collaboration with the Center of the Human Systems, TiNi Alloy has developed a tactile display consisting of a 5×6 array of tactor pins with a force of 6 g per SMA wire actuator. The response time for this display is around 100 ms.

Researchers at the University of Salford have developed a glove with tactile, contact pressure, and temperature feedback. The glove, named Teletact, is composed of a ceramic disk of PZT (lead zirconate titanate) with a 10 mm diameter and a 1 mm thickness.

There have also been wearable tactile devices embedded into a jacket. At the University of Ottawa, Professor El Saddik's team developed a haptic jacket, in which 34 pager motors (used in cell phones) are embedded (3×6 for chest and 4×4 for left upper arm) [68]. All the motors are wired to a battery-powered microcontroller, and the device is controlled through Bluetooth communication. In remote interpersonal communication, the haptic jacket provides a contact sense to a local user when a second remote user touches the local user's captured 3D image using a force feedback device. A similar jacket was developed at Philips Research Europe, where 64 uniformly distributed vibrating actuators were applied on the torso [223]. It is synchronized to a movie to present emotions such as love, enjoyment, fear, sadness, anger, anxiety, and happiness by giving distinctive tactile patterns.

4.6.3 Kinesthetic Interfaces

Kinesthetic interfaces are devices capable of feeling and manipulating objects. Kinesthesia provides humans with an awareness of the position and movement of limbs along with the associated forces that are conveyed by the sensory receptors and neural signals derived from motor commands. Kinesthetic information, such as the moving of joints, movement velocity, the contractile state of muscles controlling the joint, along with information from motor and cognitive systems, produce the perceived limb position and movement. A summary of some commercial and research kinesthetic devices is provided in Table 4.3.

Essentially, kinesthetic interfaces, or force feedback interfaces, have three main functions: (1) measuring the movements and forces exerted by a part of the human body, i.e., hand or fingers; (2) calculating the effects of these forces on objects in the virtual environment and the force response that must act on the user; and (3) applying the appropriate forces to the user. Technologies that are currently in use include electromagnetic motors, hydraulics, pneumatics, cables, and shape memory alloys. Other technologies, such as piezoelectric motors and magnetoresistive materials, have been investigated, but they are still the subject of further research and development.

A number of existing kinesthetic interfaces are now being commercially marketed or developed by research groups. The majority of these devices can be classified as exoskeleton devices, tool-based devices, thimble-based devices, or robotic graphics systems. Exoskeleton devices, such as the Force Exoskeleton ArmMaster, deliver forces to the shoulder, elbow, wrist, and finger joints. Tool-based devices deliver forces to the human hand through a knob, joystick, or pen-like object carried by a user. Some examples are the PHANToM, HapticMaster, and Impulse Engine 3000. Finally, robotic graphics systems use real objects to provide forces to the user's hand. The four force feedback configurations are illustrated in Fig. 4.11.

To evaluate the quality of force feedback systems, a set of minimum performance standards has been proposed in [313]. The authors recommend using a force output resolution of 12 bits of the maximum force output, a position resolution of 0.001 inches, and a passive friction of less than 1% of the maximum force output. The maximum force output and the range of motion is HAVE application dependent.

Table 4.3 Commercial and development kinesthetic–haptic interfaces

	Product	Feature	Sensation	Vendor
	Force Feedback Master	Desktop	Hand via joystick	EXOS, Inc. (Microsoft)
	Force Exoskeleton ArmMaster	Exoskeleton	Shoulder and elbow	EXOS, Inc. (Microsoft)
	CyberGrasp	Force-reflecting exoskeleton: five actuators, one for each finger	Resistive force feedback	Immersion Corporation
	Impulse Engine 3000	Desktop	Hand via tool handle	Immersion Corporation
	Laparoscopic Impulse Engine	Desktop	Hand via joystick	Immersion Corporation
	Interactor	Vest	Torso via vest	Aura Systems, Inc.
Commercial	Interactor Cushion	Cushion	Back via cushion	Aura Systems, Inc.
Interfaces	HapticMaster	Desktop	Hand via knob	Nissho Electronics Corporation
	Hand Exoskeleton Haptic Display	Exoskeleton	Thumb	index finger joints, palm and EXOS,Inc.(Microsoft)
	PER-Force 3 DOF	Desktop	Hand via joystick	Cybernet Systems Corporation
	PER-Force Handcontroller	Desktop	Hand via joystick	Cybernet Systems Corporation
	PHANToM	Desktop	Fingertip via Thimble	SensAble Devices, Inc.
	SAFiRE	Exoskeleton	Wrist, thumb and index finger	EXOS, Inc. (Microsoft)
	Robotic Graphics Proof-of-Concept System	Robotic graphics	Hand via tracker	Boeing Computer Services
	Force and Tactile Feedback System (FTFS)	Robotic graphics	Throttle and joystick	Computer Graphics Systems Development Corporation
	Elbow Force Feedback Display	Exoskeleton	Elbow joint	Hokkaido University
	MSR-1 Mechanical Master/Slave	Tool-based	Active limbs	MIT
	7 DOF Stylus	Tool-based	Hand via tool handle	McGill University
Research and	Force Feedback Manipulator	Desktop	Hand via joystick	Northwestern University
Development	Second Generation Rutgers Master	Thimble-based	Three fingertips and thumb	Rutgers University
	SPICE	Robotic graphics	Hand via tool handle	Suzuki Motor Corporation
	SPIDAR	Thimble-based	Thumb and index finger	Tokyo Institute of Technology
	Molecular Docking Virtual Interface	Exoskeleton	Shoulder and Elbow	University of North Carolina
	Pen-Based Force Display	Tool-based	Fingertips or pointed object	University of Washington

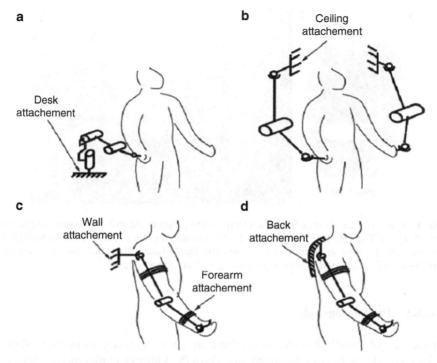

Fig. 4.11 Force feedback configurations (**a**) desk, (**b**) ceiling, (**c**) wall, and (**d**) back attachments

Other requirements include a system bandwidth of less than 50 Hz, a minimum sampling rate of 2 kHz, and a maximum latency of 1 ms. Currently, one of the few devices that meet all of these requirements is the PHANToM device; most other devices meet only a subset of these requirements. Those hardware limitations, such as the sensor's accuracy and the actuator's performance, constrain the fidelity with which haptic interactions can be simulated.

4.6.3.1   The Rotary Module

Rotary-based haptic devices can only simulate particular tasks in a one-dimensional axis (for example, opening a door with a knob that is constrained to rotate around a single axis, squeezing scissors to cut a piece of paper, or pressing a medical syringe's piston when injecting a patient). A one-DOF device measures the operator's position and applies forces to the operator along one spatial dimension only. One-DOF devices are designed for industrial control applications to perform simple operations as alternatives to traditional devices, such as mechanical switches, potentiometers, etc. One example of a current device with this type of functionality is the family of rotary modules developed by Immersion Technologies (such as the PR-3000 device shown in Fig. 4.12).

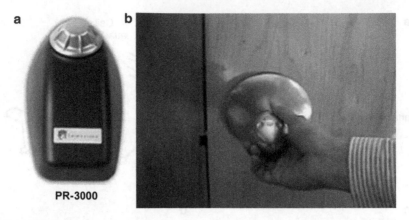

Fig. 4.12 Rotary controls from Immersion Corp. can be programmed with a wide variety of tactile sensations. (**a**) The PR-3000 Rotary Control with a Motor Actuator provides more sophisticated dynamic effects, such as springs and notches, (**b**) The "Aladdin" is a half-door with a fully functional knob, latch control, door angle sensing, and an auditory display

4.6.3.2 The Pantograph

In the case of two-DOF interactions, there are common examples throughout our daily lives, like using a mouse to interact with a PC. Other examples include devices that have been developed for the gaming industry, such as haptic steering wheels, joysticks, and game pads. Another example of a rotary controller is the two-DOF mechatronics haptic device called "Aladdin"; it includes a haptic knob with a torque and thermal display, a high quality auditory display, and sensing and actuation of other door controls. This prototype could be used to sense and actuate a door knob that provides access to an interior or remote space, such as a private or shared room, a home, or a professional building [237].

A further example of a two-DOF force feedback device is the Pantograph. It is a mechanical linkage that is connected in a particular way based on parallelograms. With this type of connection, the movement at a specified point is an amplified version of the movement of another point. It was designed for copying writing and scale diagrams. In the haptic domain, devices with two actuated degrees of freedom in the horizontal plane were initially designed with the idea of catering to the special needs of visually impaired persons [300]. The Pantograph can reconstruct interactions in real-time, creating mechanical objects with stiffness or any other physical attributes. Ramstein et al. [300] have stated that, "The response is of sufficient quality to give the users the tactile and kinesthetic sensations of rigidity, continuous outlines, sharp edges, etc." Users can explore a force field in a similar fashion to a conventional mouse.

4.6.3.3 Force Feedback Joysticks

Today, force feedback joysticks are simple, cheap, and widely used kinesthetic interfaces. Generally, they are characterized by a small number of degrees of freedom and produce moderate forces with high mechanical bandwidth. As an example, the Wingman Force 3D joystick, developed by Logitech Co. has three degrees of freedom, as well as analog buttons and switches used in gaming. The force feedback system is placed in the joystick base and consists of two DC motors as actuators. They are connected to the central handle rod through a parallel kinematics mechanism [315]; each actuator has a capstan drive and a pulley, which moves a gimbal mechanism composed of two rotating linkages. The two actuator-gimbal assemblies are perpendicular to each other to enable tilting of the central rod in four directions (right, left, front, and back). Two digital encoders, coaxial with the motor shafts, measure the tilting. The maximum applicable force is 3.3 N. Logitech also produces the Force 3D Pro joystick, which has an improved force feedback mechanism to enable more realistic interactions in gaming; however, very little details are available at this time.

4.6.3.4 The PHANToM Family

The PHANToM family was initially developed at MIT and is now marketed by SensAble Devices Inc. The Phantom is a desktop haptic interface. It has a stylus grip or a fingertip thimble with which users can reach into virtual worlds, touching and interacting with 3D objects. It measures motion along six degrees of freedom (translational and rotational dimensions) and can exert controllable forces on the user along three of those DOFs (translational only). Free motion feels smooth and comfortable because the device does not constrain motion within its workspace and because its inertia and friction are low. The relatively large dynamic range in force output, known as the ratio of the largest to the smallest displayable force, plus a good match with human resolution and bandwidth, provides enough contrast in force sensations to convincingly display impact, rigidity, texture, complex shapes, and a range of compliances. These devices are common in research laboratories because they are affordable and easy to set up and operate. Conventionally, they are well suited for tele-manipulation applications.

The Phantom Desktop interface device provides six degrees of freedom of positional sensing and three degrees of freedom of force feedback. The interface's main component is a serial feedback arm that ends with a stylus. The orientation of the stylus is passive, meaning that no torques can be applied to the user's hand. As shown in Fig. 4.14b, the interface uses three DC brushed motors (actuators) with optical encoders placed at the actuator's shaft, and a rotary potentiometer to measure the handle orientation. Transmission is achieved using cables and pulleys. The peak force of the Phantom Desktop is 6.4 N, while continuous force without overheating its actuators is only 1.7 N. The actuators are controlled by an electronics assembly that receives commands from a PC host over a parallel

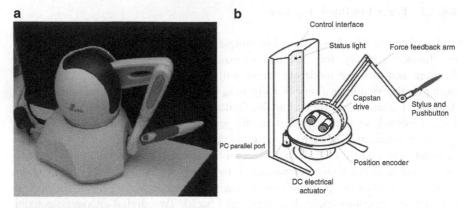

Fig. 4.13 (**a**) The Phantom Omni, from Sensable Technologies, (**b**) the force feedback system of the Phantom device

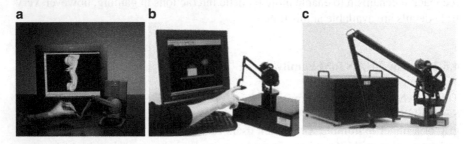

Fig. 4.14 (**a**) SensAble PHANTOM Desktop, (**b**) SensAble PHANTOM 1.0, (**c**) SensAble PHANTOM Premium 3.0

port. This electronics assembly consists of a digital-to-analog and analog-to-digital conversion card, power amplifiers for the feedback motors, conditioning electronics for the position sensors, and a status LED indicator. A snapshot of the Phantom Desktop is shown in Fig. 4.14a.

The PHANToM Omni device is similar to the Phantom device in terms of degrees of freedom, however, the maximum exertable force is 3.3 N and the device uses the IEEE-1394 FireWire port for interfacing to the PC. A photograph of the Omni device is shown in Fig. 4.13.

Due to the Phantom Desktop's limitation of output forces and its inability to generate feedback torques, the manufacturer introduced the Phantom Premium series. The Phantom Premium 1.0, shown in Fig. 4.14b, provides force feedback to a thimble that is slipped over the user's fingertip. It generates forces in three translational degrees of freedom (x, y, and z coordinates) and provides torque feedback in three rotational degrees of freedom (in the yaw, pitch, and roll directions). The torques are passed through a pre-tensioned cable transmission to a lightweight aluminum linkage that supports the thimble. The Phantom 1.5 is

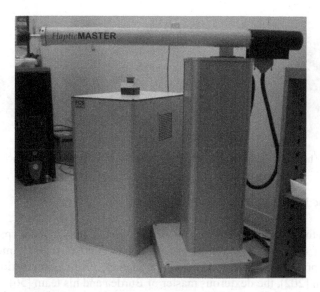

Fig. 4.15 The HapticMASTER from FCS Control Systems

essentially a later version with a 300% larger workspace to provide a range of motion approximating lower-arm movement, pivoting at the elbow. Finally, the Phantom 3 has the largest workspace and supports full-arm movement, pivoting at the shoulder. A photograph of the Phantom 3 device is shown in Fig. 4.14c.

4.6.3.5 The HapticMASTER Device

The HapticMASTER is a commercial example of a force controlled haptic interface with three degrees of freedom (translational dimensions), as shown in Fig. 4.15. This system provides the user with a clear sensation and, using a cylindrical robot arm, has the ability to closely simulate weight and force in an extended range of human tasks. The cylindrical robot can rotate around its base, move up and down, and extend its arm radially within a $0.64 \times 0.4 \times 0.36\,\mathrm{m}^3$ workspace. It recreates force with only a 0.01 N margin of error and can report the position of its end-effector to within 0.004 mm. The maximum output force is 250 N and maximum stiffness is 50,000 N m^{-1}. The control loop measures the user's forces/torques at a very high rate of 2,500 Hz, yielding a mechanical bandwidth of 10 Hz that the user can feel. This arrangement allows better simulation of hard, immovable objects, such as virtual walls. The main disadvantage is a larger apparent inertia due to the arm size. Additionally, the cost of the HapticMASTER system is relatively high due to the use of expensive force sensors and position-feedback actuators.

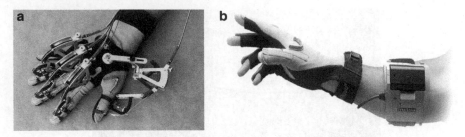

Fig. 4.16 (**a**) The CyberGrasp force feedback system; (**b**) the CyberGlove interface

4.6.3.6   The Immersion Haptic Workstation

An exoskeleton is a robotic mechanism into which human body parts can fit. The joints of the exoskeleton are aligned with the human joints. Among current exoskeleton products and prototypes are the human–machine interaction presented by Kazerooni [202], the dexterous master of Burdea and his team [56], and the arm exoskeleton system of Bergamasco and his colleagues [34]. The Immersion workstation exoskeleton comprises three components: the CyberGlove, the CyberGrasp, and the CyberForce. The CyberGlove acts as a position sensor glove that measures hand gestures, whereas the CyberGrasp and the CyberForce provide kinesthetic feedback.

The CyberGrasp glove controls simulated forces on independent fingers, rather than at the wrist, for tasks requiring high dexterity. The CyberGrasp system, shown in Fig. 4.16a, is a retrofit of the 22-sensor version of the CyberGlove. As shown in Fig. 4.16b, the CyberGlove interface unit transmits the finger position data to the CyberGrasp force control unit over a serial RS232 line. The wrist position data is then sent from a 3D magnetic tracker worn by the user to the force control unit. Eventually, the resulting 3D hand position is sent to the PC over an Ethernet line. The simulation software checks for collisions and sends back the reaction contact forces to the CyberGrasp force control unit, which then applies appropriate analog currents to the five electrical actuators. The actuator torques are transmitted to the user's fingertips through cables on a mechanical exoskeleton worn on top of the CyberGlove. As illustrated in Fig. 4.17, the mechanical exoskeleton has the role of guiding the cables using pulleys for each finger and of serving as a mechanical amplifier to increase the forces felt at the fingertip. The exoskeleton is attached to the cable guides and to the CyberGlove through finger rings, a back plate, and Velcro strips. The maximum force that can be generated at each finger is $16\,N$ within a spherical workspace of a $1\,m$ radius. Major drawbacks of the CyberGrasp are system complexity and cost, and the inability to simulate the weight and inertia of the grasped objects. Additionally, cable backlash and friction-induced hysteresis reduce the mechanical bandwidth drastically from $1\,kHz$ to around $40\,Hz$ [375].

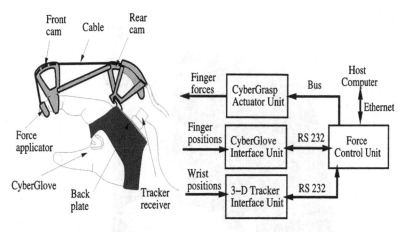

Fig. 4.17 The CyberGrasp force feedback system

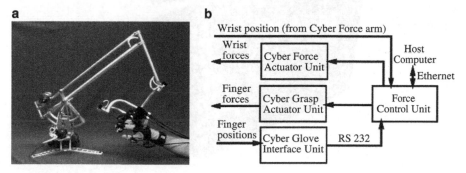

Fig. 4.18 The CyberForce force feedback system: (**a**) The outer appearance (**b**) the force feedback system

The CyberForce is an addition to the CyberGrasp for simulating the object's weight and inertia. As shown in Fig. 4.18a, it consists of a mechanical arm attached to a desk and also to the user's wrist at the CyberGrasp exoskeleton back plate (see Fig. 4.18b). The wrist and finger position data are sent to the host by the force control unit over a LAN, and the resulting contact and weight/inertia forces are sent to the CyberGrasp and CyberForce actuator units.

Using CyberForce together with CyberGrasp, you can literally "hang your hand" on a virtual steering wheel, sense weight and inertia while picking up a "heavy" virtual object, or feel the impenetrable resistance of a simulated wall. CyberForce facilitates the exploration and interaction with simulated graphical objects via the most natural interface possible – the human hand.

Fig. 4.19 The Haptic Wand System

4.6.3.7 Quanser Haptic Displays

Quanser Inc. is involved in the manufacturing of highly transparent and robust haptic systems for education, research, and industrial integration. They are involved in the heavy equipment, medical, robotic, and didactic industries [5]. Examples of existing devices are the three-DOF Planar Pantograph System [313] and the five-DOF Haptic Wand System [313]. The Haptic Wand System is a haptic device that uses a dual-pantograph arrangement, where each pantograph is driven directly by two DC motors at its shoulders and another DC motor at its waist. The device allows for three translation and two rotation (roll and pitch) degrees of freedom. It has a peak exertable force of 9 N and a peak exertable torque of 810 N mm. The force feedback workspace measures $48 \times 25 \times 45 \text{cm}^3$ and the rotational workspace measures 170° roll and 130° pitch. Figure 4.19 shows a snapshot of the Haptic Wand System.

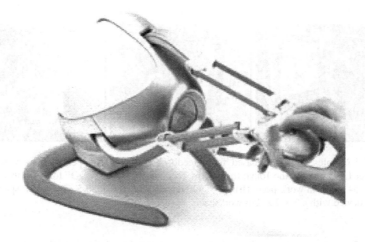

Fig. 4.20 The Novint Falcon device

4.6.3.8   The Novint Falcon

The Falcon device, developed and marketed by Novint Technologies, could be a breakthrough in the haptic device industry. It is a relatively inexpensive three-DOF device designed originally for the gaming industry. A photograph of the device is shown in Fig. 4.20. It consists of three arms extending out of the device, with one motor connected to each arm. The three arms jointly hold the device's handle, which the user grasps. The computer updates the position of the handle, and updates the currents feeding the motors at a rate of 1,000 Hz, thus providing a very realistic haptic interaction. The Falcon workspace is $12 \times 12 \times 12$ cm. The position resolution is less than 0.2 mm using optical sensors. The force torques can reach up to 5 N with constant and low friction and an update rate ranging from 800 Hz to 1 kHz [245].

4.6.4   Research and Development Efforts in Kinesthetic Haptic Displays

In this section, we discuss the work of individual research groups in the development of force feedback devices. It is worth mentioning that many of the kinesthetic interface designs originated from tele-operation and tele-manipulation projects. While we do not claim that the identified efforts are the only ones currently underway, they do form a representative set.

At the Tokyo institute of Technology in Japan, researchers in the Precision and Intelligence Laboratory are investigating the use of tension-based force feedback

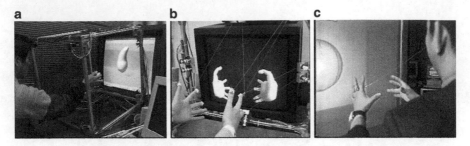

Fig. 4.21 Examples of tension-based force feedback devices developed at the Precision and Intelligence Laboratory at the Tokyo Institute of Technology. (**a**) A six DOF force feedback device with 54 × 54 × 54 cm workspace; (**b**) a 24 DOF device with 100 × 50 × 50 cm workspace; (**c**) A five DOF device with a 2 × 2 × 2 m workspace

devices. Such devices use cables that are connected to the point of contact in order to exert forces. Additionally, encoders are used to measure the length of each cable, and thus the position of the "grip" can be determined. Motors are used to create tension in the cables that propagates into the forces applied at the grip. Examples of such interfaces are shown in Fig. 4.21. The major advantages of such interfaces are the low cost, theoretically unlimited workspace, and larger amount of force that can be applied. Furthermore, the inertial efforts, and thus the accuracy of the applied forces, are much better than link-based devices due to the smaller mass of cables used in tension-based systems.

For several years, Burdea has been leading a group of researchers at Rutgers University's Center for Computer Aids for Industrial Productivity, in the development of a portable dextrous hand master. It has evolved from the Portable Dextrous Master with Force feedback (PDMFF), to the Rutgers Portable Force Feedback Master (RM-I), and finally, to the current Second Generation Rutgers master (RM-II). As an example, the RM-II comprises four custom-designed pneumatic micro-cylinders placed on an L-shaped platform, which is positioned in the user's palm and mounted on a thin leather glove [48]. The piston stroke varies from 28 to 44 mm, depending on the device setup (finger size), and can resist up to 20 N of lateral loading. The actuators are attached to three fingertips and the thumb using Velcro strips to accommodate various user hand sizes. The position sensing consists of two Hall-effect sensors mounted on the platform, one Hall-effect sensor on each cylinder, and an IR LED-phototransistor pair placed within each cylinder. Finally, a Fastrak position sensor is mounted on the back of the hand to provide wrist position and orientation.

At McGill University's Research Center for Intelligent Machines, Hayward and his group developed a six DOF (optionally seven DOF) force feedback interface device called Stylus. It is designed for use in virtual environments and tele-operation applications. Human haptic related experiments have led the researchers to decide on a workspace volume of 10 × 10 × 10 cm and an angular workspace on the order of 90° pitch and yaw, and a roll of 180° [150]. For its physical structure, the desktop

Stylus device uses grounded actuation coupled by a combination of polymeric tendon transmissions and linkages to the active end. A separate actuator pack uses conventional motors, whereas displacements and forces are measured using optical sensors.

There are still more hand master haptic devices that have been developed at many universities and research labs. In 1997, an anthropomorphic hand exoskeleton, from Vanderbilt University, was designed to prevent astronaut hand fatigue during extravehicular activities [344]. At Carnegie Mellon University, under the Robotics Institute Hand Exoskeleton Project, researchers developed an EMG Controlled Orthotic Exoskeleton for the hand, with the goal of aiding people who are unable to pinch objects between their index finger and thumb. The user would indicate their intent through the activation of another set of muscles (e.g., their biceps), and the device would supply a grasping force to the two fingers.

At the Laboratoire de Robotique de Paris, a Dextrous Hand Master (DHM) was developed by [377], which uses tendons to apply forces on each phalanx of the hand. They are able to measure 14 finger joint angles. Miniature force sensors are placed on each phalanx in order to measure cable strain and permit the implementation of force/impedance control techniques. There is also a Dextrous Hand Master developed at MIT, which consists of a carbon-fiber exoskeleton attached to an elasticized glove. The exoskeleton extends around all fingers except for the pinky finger and has a total of 16 degrees of freedom [311].

EXOS is a well-known group involved in several hand master related projects, such as the Exos Dextrous Hand Master, the Sensing and Force-Reflection Exoskeleton (SAFiRE), and the Hand Exoskeleton Haptic Display (HEHD). Another example is the mechanical design of a haptic interface for the hand from the Scuola Superiore Sant'Anna at the PERCRO Laboratory that is capable of actuating the index finger and thumb.

4.7   Final Remarks

The development of haptic devices has been influenced mainly by the accelerated growth of information technology. In addition, multimedia and interactive system requirements now necessitate more sophisticated virtual worlds incorporating the sense of touch. Thus, many research labs around the world, along with a few companies, have been leading the development of haptic devices of different sizes, actuation mechanisms, and varying applications. Medical simulators have been very influential in the development of outstandingly accurate and precise haptic devices. Meanwhile, the gaming and entertainment industries have been focused on more robust and mechanically durable devices.

Chapter 5
Computer Haptics

5.1 Introduction

Computer haptics is defined as the art and science of developing software algorithms that synthesize computer-generated forces and tactile stimuli to be displayed to the user for the perception and manipulation of virtual objects via touch. In the real world, a person can move his/her hand in order to touch an object. As soon as the fingertip touches the object, the object exerts a reaction force back against the finger to prevent it from penetrating the object. The person feels this force, along with the object's texture, through muscle and mechanoreceptors.

A real object can be represented in a virtual world by a computer-generated model, and the fingertip can be represented as a point called an "avatar." When a person moves a force feedback device with their actual finger, the corresponding point avatar mimics the movement in the virtual world. When the avatar meets the virtual object, a force similar to the real reaction force is calculated and fed back to the force feedback device to push back on the fingertip. As a result, the person feels as if they are touching a real object as depicted in Fig. 5.1. The haptic rendering algorithm is responsible for calculating the interaction force between the avatar and the virtual object. Basically, it consists of (a) a collision detection algorithm to know when the avatar meets an object and (b) a collision response algorithm to calculate the interaction force based on the collision information.

This chapter will focus primarily on the fundamental concepts of haptic rendering with some discussion about design and implementation details. Due to the vast and continuously growing interest in this exciting area of research, it is not possible to cover and cite all relevant work within the scope of this chapter.

Haptic rendering usually refers to the calculation of the interaction force between a virtual object and a user's avatar. It can be broadly categorized based on the dimension of the avatar representation as well as the number of degrees of freedom of the force feedback device used. In this chapter, we specifically discuss the haptic rendering subsystem and algorithms as well as some widely used software systems used for designing and developing HAVE applications.

A. El Saddik et al., *Haptics Technologies*, Springer Series on Touch and Haptic Systems, DOI 10.1007/978-3-642-22658-8_5, © Springer-Verlag Berlin Heidelberg 2011

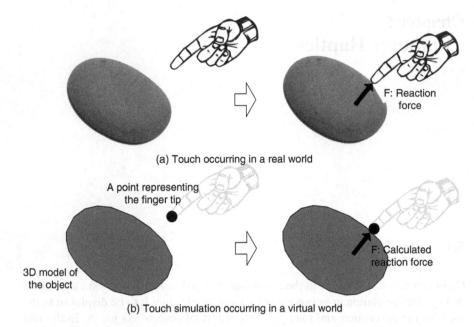

Fig. 5.1 Basic concept of touch in real and virtual worlds. (**a**) Touch occuring in a real world.
(**b**) Touch simulation occurring in a virtual world

5.2 Haptic Rendering Subsystem

The haptic rendering subsystem is the part of the HAVE system responsible for
computing interaction forces. It does this by reading a user's pose through the use
of haptic interfaces or other types of sensors and then updating the user's avatar
in the virtual environment (see Fig. 5.1). If there is no collision between the user's
avatar and an object in the virtual world, meaning the avatar is in free space, the
resultant force is zero, and there is no force feedback. If there is a collision, an
interaction force is calculated based on the penetration depth of the avatar and
the material properties of the target object. In order to simulate material properties
such as stiffness, friction, and roughness, each force component is calculated based
on the principles of physics and superposed into a resultant force that is then fed
back to the force feedback device. This force feedback loop needs to be kept in the
order of 1 kHz for stable force interaction with reasonable fidelity [53]. In addition,
the resultant force affects the status of the virtual objects by pushing and possibly
displacing or deforming them, and this state update is processed by a physics engine.

When the force feedback device is integrated with a tactile device to display
textures, the surface texture information around the collision is obtained from the
virtual object and interpreted as device actuation commands for tactile devices. The
actuation commands will vary since they are based on display algorithms specific to
each tactile device. This is shown in Fig. 5.2.

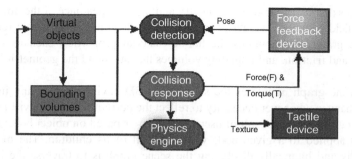

Fig. 5.2 Flow of information for tactile and force-feedback interaction

In addition to interaction forces with virtual objects, resultant forces can be affected by environmental force effects. For example, if there is a magnetic field in the virtual world, the magnetic force imposed on the avatar needs to be superposed to the resultant force. Force fields, such as magnetic and gravitational forces, can be modeled around basic physical rules and stored in the representation of the virtual environment.

Algorithms for collision detection and response vary based on the representation methods of the virtual environments. In the following subsections, three-DOF haptic rendering algorithms are introduced according to the polygon-based representation method, which is commonly used in haptics applications. Other haptic rendering algorithms that are based on different representation methods and multi-DOF haptic rendering algorithms will be briefly explained later and are also found in [227].

5.3 Polygon-Based Representation and Scene Graph

The basic building block of a virtual environment is a 3D model. Therefore, the virtual environment can be seen as a composition of 3D object models that populate the landscape. The polygon-based representation is one of the many representational methods for 3D objects. The basic object used in the polygonal modeling is the vertex, which is simply a point in 3D space. Multiple vertices connected together in one plane form a polygon or a face. A group of polygons that are connected to each other by shared vertices is generally referred to as an element or a mesh. Due to its simplicity, the polygon-based representation has been frequently used in 3D computer graphics as well as in haptic rendering.

A three-dimensional environment is defined as a set of geometric primitives and their display properties. A scene graph is a hierarchical structure (tree) of nodes used to define the geometry and the display properties of both graphical and haptical objects [57]. Each node in the hierarchy represents either the geometry or a property of the geometry, such as color, stiffness, or location. Consequently, the scene graph stores information about the geometry of the object, its appearance, its behavior (procedures that can change the geometry and/or the appearance of the object), and its force field properties (gravity, magnetic field magnitude). In haptic rendering, the scene graph also contains the haptic material properties, such as stiffness, damping, and friction, and bounding volumes that surround the geometries in each node.

The scene graph maintains the state of the 3D environment at any instant in time. It changes whenever necessary to reflect the current state of the virtual world. Such changes may be a result of user interaction or based on object behavior. Any operation applied to a given node will affect all of its children. The method of graphically and haptically displaying the scene graph is to traverse the tree in a predefined manner, thus displaying the geometric primitives based on their specified properties.

5.4   Collision Detection Techniques and Bounding Volumes

In three-DOF haptic rendering algorithms, a user's avatar is represented as a point in the 3D space. This avatar is called the Haptic Interface Point (HIP). Generally, a collision occurs when the HIP hits the surface of a virtual object. However, in a discrete system such as a virtual environment, the HIP position is sampled between specific time intervals, so the HIP might penetrate the object's surface in error. This is described in Fig. 5.3a, where the black dot represents the HIP, and t_{n-1} and t_n depict times at the $(n-1)$th and (n)th samples, respectively. In the early stages of haptic rendering algorithm development, a technique called Vector Field Methods was used [246]. To use this technique, the internal volume of a virtual object was divided into subvolumes associated with the penetrated surface to test whether the HIP was inside the object. However, this method has proven difficult in calculating the proper forces exerted, and it only works for simple geometries because the subvolumes need to be constructed by hand. In order to avoid these difficulties, a line segment representing an approximate path that the HIP follows between the $(n-1)$th and (n)th sampling times, as depicted in Fig. 5.3b, is considered in

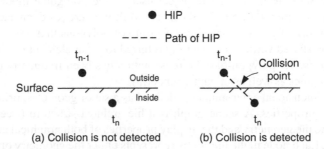

(a) Collision is not detected   (b) Collision is detected

Fig. 5.3 The collision detection between an object's surface and the line segment connecting the previous HIP and the current HIP positions. (**a**) Collision is not detected. (**b**) Collision is detected

the collision detection algorithm. By doing this, the extra process to construct the internal volume is avoided, and the collided surface and the collision point can be identified inherently.

It takes considerable time, compared to the high update rate requirement for haptic rendering (1 kHz or 1 ms) to test for the collision of a line segment with all the polygons in the virtual environment. It is even more time consuming when the complexity of the virtual object increases.

One technique for accelerating the collision detection process uses bounding volumes. The primary goal of using the bounding volumes is to get fast rejection tests. In the bounding volume intersection test, simple primitives are used as bounding volumes for complex-shaped objects. If the bounding volumes of two objects do not intersect, then the objects do not. If they do intersect, then further testing is required. The bounding volume approach uses the cheap static intersection collision test. An example of this kind of volume bounding technique is called the Axis Aligned Bounded Box (AABB) technique [135]. The overlap test using AABB is far less expensive than the line segment-polygon intersection test.

When using a primitive as a bounding volume, three things must be considered:

- How well does the bounding volume fit with the underlying geometry?
- What is the cost in updating the bounding volume if the object is moving?
- What is the cost of the bounding volume intersection test?

To explain the effects of the bounding volume geometry, we will discuss four implementations of bounding volumes, namely the sphere volume, AABB, the Oriented Bounding Box (OBB), and a more complicated OBB implementation, all of which are shown in Fig. 5.4.

The sphere volume method uses a sphere bounding volume to encompass the target geometry. The sphere can contain a lot of unoccupied space, so many bounding volumes could overlap while their geometry inside does not. This increases the number of unnecessary collision tests on the internal geometries. Furthermore, as

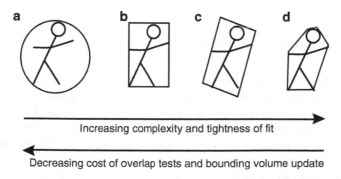

Fig. 5.4 (a) Sphere volume, (b) axis aligned bounding box, (c) oriented bounding box, and (d) more complex OBB

the object moves, the bounding volume needs to be updated. While this is trivial in the case of the sphere, the OBB update depends on the motion of the object. If the motion is rigid (rotation and translation only), the transformation must simply be applied to the bounding volume as well. If, on the other hand, the geometry is deforming, the update is far more complex. In addition, overlap testing for the sphere and AABB methods is relatively simple when compared to the OBB method. From these examples, one can see that there is always a tradeoff between the quality of the fit, the cost of updating, and the cost of intersecting a given bounding volume. Finding an optimal balance between certain types of bounding volumes is an ongoing topic for debate and research.

These bounding volumes can also have a hierarchy in a scene graph. A bounding volume at a node can surround its own geometry as well as the geometries of its child nodes. By having a hierarchy, the line segment is first tested with the bounding volume of the root node and then traverses to the bounding volume of a leaf node. If there is a collision with the last bounding volume, collision is tested with the geometry in that leaf node.

5.5 Penetration Depth (Penalty-Based Approach) and Collision Response

Once a collision is detected, force-response algorithms are triggered, and interaction forces between the avatar and virtual object are computed. Due to the mechanical compliance of haptic interfaces and the discrete sampling of the HIP, the avatar often penetrates the virtual object and maintains the penetration depth during continuous touching. In the force response calculation, the restoring force is calculated using the penetration depth, which is the distance from the HIP to the closest surface, based on Hooke's Law. This is commonly known as the penalty-based method.

5.5.1 Proxy-Based Approach

The penalty-based method (also called the penetration depth method) has two issues. First, the force abruptly changes at the edge of the polygonal surface when an adjacent surface is connected with a significant angle difference. For example, as shown in Fig. 5.5a, while the HIP moves from a given position at time t_{n-2} toward a new position a time t_{n-1}, the closest surface is surface A. The resultant force is upward, and a user can feel surface A as expected, however, when the HIP goes closer to surface B, it becomes the closest surface, and the force direction changes significantly so that the user feels like they are being pulled to the right side. Second, when a user touches a thin object from a given position at t_{n-2}, and

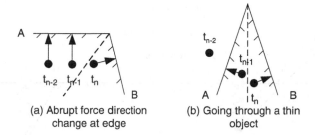

(a) Abrupt force direction   (b) Going through a thin
change at edge   object

Fig. 5.5 Two noticeable defects in the penalty-based method (the *black dot* represents the HIP).
(**a**) Abrupt force direction change at edge. (**b**) Going through a thin object

the HIP penetrates the surface A at t_{n-1}, as shown in Fig. 5.5b, they feel the reaction
force pushing toward the outside of surface A, which is the closest surface to the
HIP. While the user continues to explore surface A, they may cross the middle line
of the two surfaces and cause the closest surface to abruptly become surface B. As
a result, the user HIP is forced out of the thin object, but on the opposite side of
the object. Also, if the object is too thin, the user might just go through it without
feeling any surface at all.

Those errors described in Fig. 5.5 are induced due to the fact that the HIP
cannot be prevented from penetrating surfaces. In this context, a proxy-based
method was introduced and became an essential concept in haptic rendering
algorithms [320, 410]. A proxy, also called a god-object, is an ideal HIP that has
no mass. The proxy cannot penetrate any surfaces and is connected to the HIP
with an ideal spring that can elongate from 0 to infinite. In free space, the proxy
coincides with the HIP. When the HIP penetrates a surface, the proxy collides with
the surface without penetrating and continues sliding to the closest position that
minimizes the potential energy in the spring. At the same time, the stiffness of the
ideal spring becomes the stiffness of the surface that the proxy is in contact with.
The restoring force is calculated based on Hook's Law, using the distance between
the HIP and the proxy. Therefore, determining the proxy position can be considered
the same as the process of calculating the force response. For example, when an
HIP penetrates a surface, as shown in Fig. 5.6a, the proxy is at location of the HIP at
the previous sampling time (i), but it moves toward the HIP due to the ideal spring,
and ends up colliding with the surface. This collision position (ii) can be obtained
by testing for a collision between the surface and a line segment (represented by a
dashed line in the figure) that connects the previous proxy to the current HIP. After
the collision, the collided surface is set as the active surface, and the proxy slides
to the position on that surface that is closest to the current HIP (iii). The closest
position can be calculated by projecting the current HIP onto the plane of the active
surface. The final resultant force can be obtained by setting a spring with the active
surface's stiffness coefficient. As the HIP moves over the next time interval, the
same process is performed, and the proxy position is updated to position (iv) in
Fig. 5.6b. However, when the current HIP goes over perpendicular boundaries of

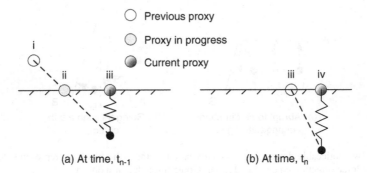

Fig. 5.6 Updating the proxy position after a collision occurs. (**a**) At time, t_{n-1}. (**b**) At time, t_n

the active surface, the proxy can float in the air or go through an adjacent surface, resulting in distorted forces.

In the case of a convex portion, as shown in Fig. 5.7a, the proxy moves from (i) to (ii), where (ii) is the closest position to the current HIP on the plane of the active surface A. However, the resultant proxy position is located beyond the active surface and is floating in the air. This may induce a small distortion but is almost imperceptible considering the small time period and the distance traversed during the haptic rendering process. Nevertheless, this distortion can be easily removed by repeating the collision detection process until there is no collision with any surfaces other than the active one. For example, as shown in Fig. 5.7b, the proxy slides from (iii) to (iv), and another collision detection is performed where the adjacent surface B becomes active. Finally, the correct proxy position is obtained at (v). However, these additional collision detection processes can make the whole haptic rendering process slow, especially in the case of very small polygons when the HIP passes many surfaces in one haptic rendering loop. Collision detection is the most time-consuming factor, and collisions need to be examined on each surface that the proxy passes by.

When touching a concave portion of an object composed of surfaces A and B, as shown in Fig. 5.7c, the proxy slides from (i) to the closest point on the plane of the active surface A to the HIP (ii). However, since surface B is not considered, the proxy at (ii) has already penetrated the neighboring surface B and is inside the object. This problem can be fixed by testing for collisions between the neighboring surface and the line segment connecting the previous proxy (iii) and the new candidate proxy (iv), as illustrated in Fig. 5.7d. As a result, the line segment collides with the neighboring surface B, which becomes an active surface. Note that surface A is still active because the line segment between the HIP and the new proxy collides with surface A. This additional process needs to be iterated until all neighboring surfaces are examined. Since the new proxy can be in contact with many neighboring surfaces, there could be more than one active surface in a concave portion of a 3D object.

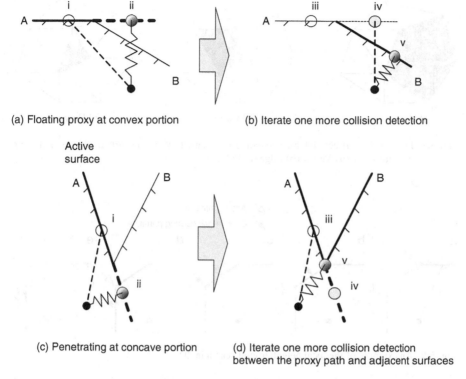

Fig. 5.7 Failed collision detection at a concave portion of a virtual object (**a**), distorted proxy position at a convex portion of a virtual object (**c**), and their solutions (**b** and **d**). (**a**) Floating proxy at convex portion. (**b**) Iterate one more collision detection. (**c**) Penetrating at concave portion. (**d**) Iteration one more collision detection between the proxy path and adjacent surfcaces

5.5.2 Local Neighbor Search

A local neighborhood search (LNS) can help reduce the number of collision detections by sacrificing more memory storage [168]. In the LNS method, a virtual object is considered as consisting of primitives such as vertices, edges, and polygons. Each primitive includes both its own geometric information as well as the information of neighboring primitives. For instance, when considering triangular meshes, a vertex has information for that point, but additionally includes neighboring edge and polygon information. For example, in Fig. 5.8a, the vertex information also contains the five neighboring edges, each represented with a thick line, and the five neighboring triangles filled with different patterns. An edge has its own line information, but also includes information on the two neighboring vertices represented by dots, and the two neighboring triangles (Fig. 5.8b). A triangle face has its own surface information and includes three neighboring edges and three vertices (Fig. 5.8c). Through preprocessing, all the neighborhood

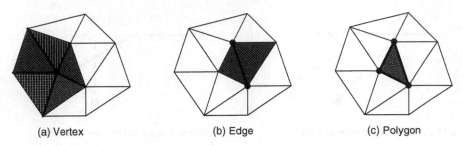

(a) Vertex   (b) Edge   (c) Polygon

Fig. 5.8 Primitives (vertices, edges, and polygons) consisting of 3D virtual objects and their neighboring primitives. (**a**) Vertex (**b**) Edge (**c**) Polygon

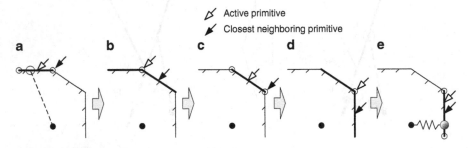

Fig. 5.9 An example of the local neighborhood search in 2D

information for each primitive can be obtained and stored within the geometric representation.

Figure 5.9 illustrates the LNS procedure in 2D space for easier comprehension. The LNS method starts with collision detection. When there is a collision with a polygon, that polygon becomes active and the collision point is stored. The active polygon is indicated with a hollow arrow in Fig. 5.9. Whenever a primitive is made active, the method compares distances for the neighboring primitives to find the one that is closest to the HIP. If the current active primitive is further than one of the neighboring primitives, the active primitive is replaced by the new closest primitive, and the proxy position is set to the point on the new active primitive that is closest to the HIP. For example, in Fig. 5.9a, the active edge has two neighboring vertices. The right side vertex is closest to the HIP, so it will be set active. This process of checking neighboring primitives is repeated until the current active primitive is the closest one to the HIP. As shown in Fig. 5.9a–d, the active primitive is repeatedly updated to the closer primitives. Finally, in Fig. 5.9e, the current active primitive is closer to the HIP than any other neighboring primitives, and the final proxy position is determined on the current active primitive unless the HIP is outside of the geometry. The final restoring force is calculated if the HIP is inside the geometry.

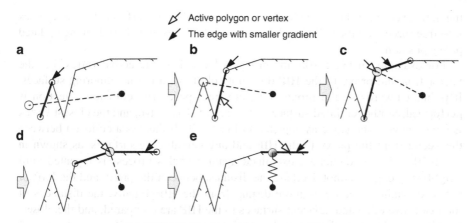

Fig. 5.10 An example of local neighborhood search on vertices in 2D

5.5.3 Local Neighbor Search on Vertices

This neighborhood search can be accelerated by restricting the position of the proxy to vertices only [386]. For this method, preprocessing is again conducted to build up neighborhood information. In the first stage, the collided surface is set active, and the vertex from that active surface that is closest to the HIP becomes the start point, as shown in Fig. 5.10a. Instead of checking the distance from each neighboring vertex to the HIP, edge gradients are compared. The edge gradient indicates how fast the proxy approaches the HIP on the neighboring edges. It is calculated by performing the dot products[1] between two values:

- The normalized edge vectors from the active vertex to neighboring vertices
- The normalized vector from the HIP to the active vertex

By choosing the vertex that corresponds to the smallest edge gradient, the fastest route from the proxy toward the HIP can be obtained. For example, in Fig. 5.10c, the active vertex is indicated by the hollow arrow, and the right side edge is selected as a path because it decreases the distance from the proxy to the HIP the fastest. Once the edge with the smallest gradient is chosen, the vertex at the other end on the edge is set active. After selecting the next vertex as a new proxy position, it is checked whether the proxy will exit the object or move onto another vertex. This can be done by calculating the dot product of the surface normal of neighboring vertices and the vector from the vertex to the HIP. In other words, it checks whether the vector connecting the proxy to the HIP collides with neighboring surfaces around

---

[1]Dot product: is an algebraic operation that takes two equal-length sequences of numbers (usually coordinate vectors) and returns a single number obtained by multiplying corresponding entries and adding up those products. Source: Wikipedia.

the active vertex. If there is no collision with neighboring surfaces, the proxy goes into free space, and the collision detection process is resumed based on the updated proxy position.

For example, in Fig. 5.10b, the proxy is located on the active vertex, but the vector from the proxy to the HIP does not collide with the neighboring surfaces. It points outwards so the proxy goes into free space. The collision detection is performed, another collided surface is then selected as active, and the closest vertex is consecutively set active as depicted in Fig. 5.10b. If there is a collision between the vector from the proxy to the HIP and any neighboring surfaces, as shown in Fig. 5.10c, edge gradients are examined again, and this process is repeated until neighboring edges cannot decrease the distance between the proxy and the HIP, or the next candidate vertex is more distant from the HIP. Finally, the distances of the last active edge and adjacent surfaces to the HIP are compared, and the closest primitive is selected to determine the current proxy position on it.

For example, in Fig. 5.10d, the right side edge decreases the distance between the proxy and the HIP faster, and thus is selected as active. Consequently, the right side vertex is set active. However, the previous vertex is closer than the newly chosen one, so the new proxy position is determined by comparing its distance to the HIP on the edge between the two vertices and on the adjacent surfaces.

5.5.4   Local Neighbor Search on Correct Path

The main goal of the two previously described neighborhood search algorithms is to find the new proxy position closest to the HIP on the surface of a virtual object as fast as possible. These two algorithms do not consider the path of the proxy between the original and new proxy positions. Figure 5.11a, b show the two resultant paths using the above-described algorithms on a flat object, under which the HIP is positioned. Although the new proxy positions obtained are the same, the proxy paths during the calculation are different from the ideal path depicted in Fig. 5.11c. This does not affect the restoring force when there is no force effect on the surface, such as friction, however, when friction is a factor, the incorrect paths can lead to a wrong resultant force direction.

As an example, when a friction cone algorithm (a generalized method to compute friction forces on a simulated haptic object [254]) is applied for friction calculation, the proxy starts to move from the previous position toward the HIP and stops due to friction at the boundary of the friction cone's projected circle. A detailed explanation can be found in [148, 254]. In Fig. 5.12a, the proxy moves on the primitives that are successively closer to the HIP than all other neighboring primitives, but at the last stage the proxy stops on the edge of the projected friction cone, and the resultant force is calculated based on this proxy position. The neighbor search on vertices calculates a different proxy path and a different resultant force, as shown in Fig. 5.12b. However, the correct dragging friction force on the flat surface should have an opposite direction to the hand movement, which approximately points

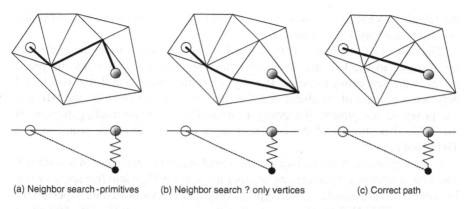

(a) Neighbor search-primitives   (b) Neighbor search ? only vertices   (c) Correct path

Fig. 5.11 Proxy paths on the flat surface, which are correct and obtained by neighbor search algorithms (*top view* and *side view*). (**a**) Neighbor search – primitives. (**b**) Neighbor search – only vertices. (**c**) Correct path

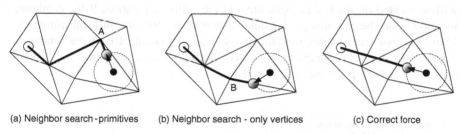

(a) Neighbor search-primitives   (b) Neighbor search - only vertices   (c) Correct force

Fig. 5.12 The resultant proxy positions based on the friction cone algorithm for each proxy path. (**a**) Neighbor search – primitives (**b**) Neighbor search – only vertices (**c**) Correct force

from the HIP toward the previous proxy position. This correct friction force can be calculated by obtaining the correct (ideal) proxy path. When the path is correct, the proxy stops at the friction cone with a resultant force opposite to the user's movement, as shown in Fig. 5.12c.

5.5.5   Triangular Mesh Modeling

In order to overcome the above-mentioned problem, the neighbor search algorithm can be modified to obtain a correct proxy path [67]. The explanation of the algorithm is based on a triangular mesh representation. In order to simplify the haptic rendering process, three types of primitives are defined: TRIANGLE, EDGE, and VERTEX. As before, each primitive contains its geometric information as well as each neighbor primitive's information. The TRIANGLE has three vertices and two normal vectors of opposite directions as geometric information and three EDGEs as neighbors. An EDGE's information has two vertices and two TRIANGLEs.

And finally, a VERTEX's information has a vertex and a certain number of EDGEs that share this vertex as an end point.

The core of this algorithm is to search for a new proxy location that minimizes the distance to the HIP and eventually finds the shortest path along the proxy traces to the new proxy location. The algorithm starts when a collision is detected between a triangle of an object and the line segment that connects the HIP and the proxy. Consequently, the proxy is moved onto the obstructing triangle at the collided position, and the current triangle is set to an active primitive as a TRIANGLE.

Once a primitive is marked active, the neighborhood search algorithm is started. The first procedure is to determine whether the proxy will go into free space or not. In order to avoid redundant, overlapping computation, this procedure is performed only at the TRIANGLE. This means that the proxy can only go into free space through a TRIANGLE. Then, the algorithm computes the candidate of the new proxy location. If the candidate location is not on the primitive and goes over any neighbor, the active primitive is updated to the neighbor primitive, and the proxy will be located on the updated primitive at the collision position. If the candidate location is on the primitive, it becomes a new proxy location in local minimum. These processes are repeated until the proxy location is obtained at a local minimum. Figure 5.13 depicts a complete flow chart outlining the algorithm. The detailed procedure on each primitive is followed.

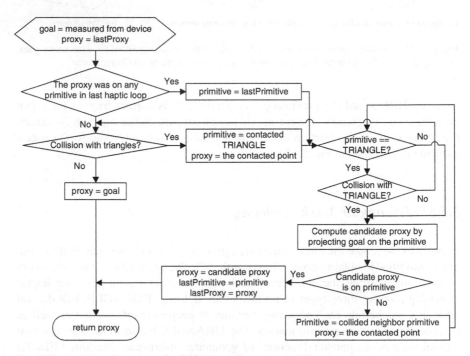

Fig. 5.13 Complete flow chart of neighbor search on correct path.

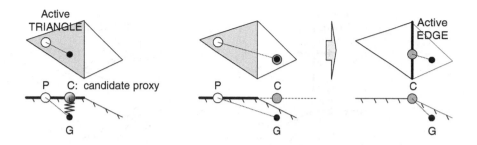

(a)Candidate proxy on TRIANGLE (b) Candidate proxy outside TRIANGLE and transition to EDGE

Fig. 5.14 Updating candidate proxy and transition on TRIANGLE. (**a**) Candidate proxy on TRIANGLE. (**b**) Candidate proxy outside TRIANGLE

At the TRIANGLE, whether the proxy goes into free space or not is checked by computing the dot product between the vector from the proxy to the goal and the normal of the TRIANGLE. If it is larger than zero, the proxy goes into free space, and the collision detection is performed again. When a collision is detected, a candidate proxy location is determined by projecting the HIP on the plane that includes TRIANGLE. If the candidate proxy location is inside TRIANGLE, it becomes the new proxy location in local minimum as shown Fig. 5.14a. Otherwise, the proxy should eventually stop on an EDGE that the proxy path collides with, as shown in Fig. 5.14b. This procedure is performed by checking the collision between the line segment PC, and each of the three neighbor EDGEs. The colliding EDGE is set to active.

At an EDGE, the proxy can go onto one of the four neighbors, two TRIANGLEs and two VERTEXs. First, a check is performed to evaluate whether the proxy can go on the TRIANGLEs. Each EDGE has two normal vectors, m1 and m2, as shown in Fig. 5.15a. They point toward each neighboring TRIANGLE that is perpendicular to the EDGE and aligned in opposite direction. In order to determine which TRIANGLE decreases the distance from the proxy to the HIP at a faster rate, the dot products (distance gradients) of the normals and the normalized vectors from the HIP to the proxy are compared. The TRIANGLE that has the smaller gradient is set as active.

Figure 5.15b shows that if the two gradients are positive, it is recognized that two TRIANGLEs cannot decrease the distance, so the proxy slides along the EDGE. The candidate proxy location is obtained by projecting the goal onto the line that goes through the EDGE. If the candidate is on the EDGE, the candidate location becomes a new proxy location in local minimum. Otherwise, the VERTEX to which the candidate moves will be active, and the proxy becomes the VERTEX.

At a VERTEX, the distance gradients for each EDGE that has the VERTEX as an end point are compared using the previous neighborhood search algorithm. The

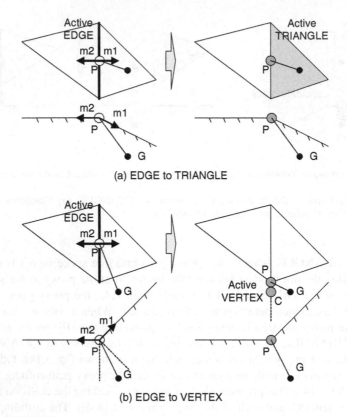

(a) EDGE to TRIANGLE

(b) EDGE to VERTEX

Fig. 5.15 Updating candidate proxy and transition on TRIANGLE and VERTEX on EDGE. (a) EDGE to TRIANGLE. (b) EDGE to VERTEX

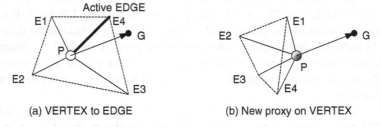

(a) VERTEX to EDGE   (b) New proxy on VERTEX

Fig. 5.16 Updating candidate proxy and transition on EDGE and on VERTEX. (a) VERTEX to EDGE. (b) New proxy on VERTEX

EDGE that has the smallest gradient becomes active, as shown in Fig. 5.16a. If all gradients are positive, the point of the VERTEX becomes a new proxy location in local minimum, as shown in Fig. 5.16b.

5.6 Haptic Rendering of Surface Properties

The previous subsections dealt with how interaction forces can present macro-geometric object information such as shape. This subsection now focuses on micro-geometric details that act as obstructions when two surfaces slide against each other and generate forces tangential to the surface and opposite to the motion. We present several algorithms that can render virtual objects' haptic textures and friction properties. Initial efforts used simple empirical models as bases for simple frictional models with 3-DOF [320, 410].

Over time, researchers from outside the haptic community developed many models to render friction with higher accuracy. Some examples are the Karnopp model for modeling stick-slip friction, the Bristle model, and the reset integrator model. Since there is always a trade-off between accuracy and delay, which is a critical factor for real-time applications, researchers worked to develop even more accurate algorithms for friction. For example, Hayward and Armstrong introduced a time-free, drift-free, multi-dimensional model of friction [151]. The computational friction model provides several advantages, including autonomy for nonuniform sampling, robustness to noise, and extensibility to 2D and 3D motions.

As for texture rendering, researchers have proposed many techniques for rendering forces to replicate texture, many of which are inspired from techniques in computer graphics. In computer graphics, realistic texture is achieved by projecting a bitmap image onto the rendered surfaces. The same can be done haptically. The work in [260] proposed haptic texture mapping for 2D scenes, which was later extended in [320] to 3D scenes.

Moreover, mathematical functions have also been used to create synthetic patterns. For instance, Basdogan et al. [28] and Costa and Cutkosky [87] investigated the use of fractals for modeling natural textures. They focused on the display of roughness using models that produce surface profiles identified by two parameters: root mean square amplitude and fractal dimension. The perceived roughness is related to variations in these two parameters when interacting with the surfaces through a haptic display. Other work has examined the use of stochastic methods for texture display [349]. In this work, the surface contact force is decomposed into traditional, rigid body contact normal (constraint) and lateral (friction) components.

5.7 Haptic Rendering for Other Representation Methods

Polygon-based representation does not provide any information about the internal volume of a virtual object. This knowledge is very important in applications such as medical simulations. An example would be the simulation of an interactive cutting operation for a human organ or for skin with bone underneath. In such cases, volumetric representations (such as voxels) can be used. A voxel-based object is represented as a 3D rectilinear array of volume elements called voxels, each specifying a large number of physical properties such as density, stiffness, and viscosity. Although each voxel does not have surface property information, users can feel a gradient force that is obtained from the intensity values used in mathematical functions [20, 328]. A proxy-based algorithm was also introduced to enable a user to touch the surface of the voxel model [184]. In the context of data representation in voxel-based models, the haptic information is contained in each voxel, along with the intensity values. In some research, each voxel has its own surface properties directly beside the intensity values [252]. More details about volumetric representation can be found in [203, 218]. Its challenges include a significant degradation in haptic rendering accuracy and memory ineffi-ciency [14].

Implicit representation uses geometric primitives (spheres, cones, cylinders, etc.) that are defined through mathematical expressions and wrapped around geometric models for force rendering [203]. The haptic properties are implicitly assigned to the surfaces. Implicit representation provides several advantages [14]. First, it enables faster and easier collision detection because a simple point inclusion function can be used to calculate collisions between objects and points in space. Second, the tangent to a surface can be easily calculated in order to display surface properties. Finally, several arithmetic operations, such as addition, subtraction, and concatenation, can be applied to make more complex objects. More details about implicit surface representations for haptic rendering can be found in [328]. With all of these advantages, there still remains the issue of determining which point on the surface should be used to model the collision interaction forces. In real-time scenarios, where quick-and-dirty rendering is required, representa-tion methodologies such as the Non-Uniform Rational B-Spline (NURBS) and Bézier patch have been widely used [372]. NURBS surfaces, typically used in graphics, have the advantage of compactness, embedded smoothness, and exact computation of surface tangents and normals [371]. The NURBS representation for haptic properties is the same as for implicit surfaces. Due to their computational efficiencies, these representation methods are widely used in deformable body simulations, such as sculpturing and surgery training [122, 203]. However, these methods have problems describing sharp edges when compared to the polygon-based method.

Implicit and NURBS representations are restricted in their use because of their limitations in describing complex objects. The most common representation methods are polygon based for general purposes and voxel based for specific medical simulations. However, these representation methods are not suitable for broadcasting or multicasting in multimedia systems where multimedia contents are instantly available because downloading massive quantities of data takes significant time. Consequently, the depth image-based representation (DIBR), which was originally proposed in [183], can be used in these cases. DIBR uses two images for each video frame: the RGB image and the depth image. Several advantages of using DIBR have been pinpointed in [225]. First, existing methods of image processing and compression can be applied to DIBR due to its simple and regular structure. Second, real-world objects and scenes can be rendered without the need for millions of polygons and expensive computations. Finally, the rendering time remains constant and independent from the complexity of the scene since it is proportional to the number of pixels of the captured images. Cha et al. [70] adopted this representation to enable viewers to touch a 3D video scene. However, since DIBR did not contain any haptic information, it could not provide a rich interaction force feeling. In addition, the haptic rendering algorithm did not support haptic texture rendering. They later adapted their approach to incorporate haptic surface properties into DIBR, and eventually introduced Depth Image-Based Haptic Rendering (DIBHR), a haptic rendering algorithm that supports haptic texture rendering [68].

5.8 Haptic Rendering of More Than 3-DOF

Although the 3-DOF point-based interaction paradigm has provided a convenient tool-based interaction that allows users to grab a tool-shaped end-effector and interact with a virtual environment with a tool-tip, natural and dextrous interaction, like that which we experience in real life, is lacking. One of the promising interaction paradigms is 6-DOF tool-based haptic interaction, which increases torque capabilities to 3-DOF. While the 3-DOF haptic rendering allows a point to interact with an object and provide axial reaction forces only, 6-DOF leads to object–object interactions that present reaction forces and torque at the same time. In [252], a voxel-based representation of a complex environment of static rigid objects was used to implement a 6-DOF interaction. This approach uses short-range force fields surrounding the static objects. The force fields repel the manipulated object to try and maintain a voxel-scale minimum separation distance that prevents exact surface interpenetration. This model is suitable for applications that can tolerate voxel-scale minimum separation distances, such as assembly task simulations.

In addition, the haptic display of complex object–object interactions has been simulated and demonstrated in [276]. In this work, a multi-resolution hierarchy is constructed and used as a bounding volume hierarchy for time-critical contact force computation in haptic rendering.

Another algorithm for the haptic display of moderately complex polygonal models using a 6-DOF force feedback device is presented in [140]. The solution uses incremental algorithms to determine when there is contact between convex primitives. This contact information is then used to calculate the restoring forces and torques, and thereby generate a sense of virtual touch. To speed up the computation, several concepts, such as geometric locality, temporal coherence, and predictive methods are exploited.

Another haptic rendering algorithm for arbitrary polygonal models using 6-DOF haptic interfaces is described in [194]. This approach finds the local minimum distances between polygonal models using spatialized normal cone hierarchies [195].

The haptic rendering process computes forces and torques on the moving model based on these local minimum distances. To provide higher haptic interaction rates on more complex scenes, Johnson et al. [194] propose a global search for local minimum distances to provide repulsive forces between models. The global search continuously adds and deletes local minimum distance pairs that are being updated by the local search.

Another way to provide more dexterity is to consider each finger as an interaction point. This can allow actions such as pinching and grabbing. Barbagli et al. [24] simulated a 4-DOF interaction through soft-finger contact. In order to simulate a soft finger contact, a 4-DOF proxy was used. Three of these degrees of freedom describe the position of the contact point when touching a virtual object, while the fourth variable describes the relative angular motion between the two soft finger avatars and a virtual object. Harwin and Melder applied their 3-DOF friction cone algorithm for two fingers independently [148]. In other words, they simply performed the 3-DOF haptic rendering algorithm for each point. In their system, users fit each of their two fingers to 3-DOF haptic devices and interact with virtual objects. By doing this, the user can grab and move objects.

5.9 Control Methods for Haptic Systems

Force response algorithms compute the ideal interaction forces between the haptic interface avatar and the virtual environment. The force's exact value, often called the desired impedance, cannot be directly applied to the user due to many haptic device technology limitations. First, since haptic interfaces can only exert forces with a limited magnitude, and not equally in all directions, force saturation must be avoided as it would lead to discontinuous application of forces to the user, and eventually, instability. Second, haptic devices are not ideal force transformers. The friction, inertia, and backlash present in most haptic devices prevent ideal performance. Third, the discrete-time nature of haptic-rendering algorithms is considered the major challenge that prevents continuous user operation. Finally, haptic device position sensors have a finite resolution, so determining the contact points and time always results in quantization errors. In considering these limitations, control algorithms must command the haptic device in such a way that they minimize the error between ideal and applicable forces.

A brief description of control architectures used in haptic systems has been presented in [129]. The objectives are to provide the kinesthetic constraint of the virtual environment and to improve the transparency of the device by decreasing the inertia felt by the user in unconstrained movement. Essentially, these architectures compute the transfer function that relates the force exerted by the user to the displacement of the haptic interface. The classification of these strategies is made in accordance with the interface inertia and the compensation method, and is shown in Fig. 5.17.

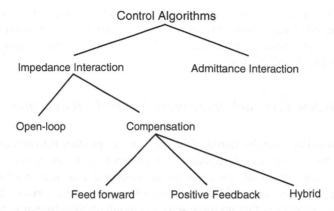

Fig. 5.17 Classification of control algorithms

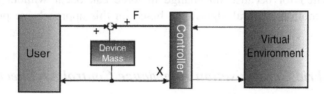

Fig. 5.18 Impedance control architecture

5.9.1 Impedance Control Architecture

This method is the most widespread architecture for haptic interfaces with small inertia and is usually referred to as the impedance control architecture. The contact forces are computed using impedance causality and are restored via negative feedback. The total impedance felt by the user is the sum of the user's arm dynamics, the interface impedance, and the environmental impedance. This configuration has been deeply analyzed in order to control the amount of excessive energy the system generates and to maintain stability. Some passivity conditions and experimental stability results can be found in [55,80,83,97,128,130,259]. This concept is depicted in Fig. 5.18, where X represents the rendered position and F corresponds to the force read by the haptic device.

5.9.2 Feed-Forward Impedance Control Architecture

This strategy has been developed to compensate for the dynamic behavior of moderate inertia interfaces while maintaining the impedance interaction. The inertia felt by the user can be decreased by including a force feed-forward that helps the operator move the interface. The inertia could be arbitrarily reduced by simply setting the force gain as large as possible. However, this can make the system unstable since the force sensor represents a source of instabilities. Another drawback is that the force sensor is an expensive device. Many of these algorithms are described in [34, 37, 62, 77, 120].

5.9.3 Positive Feedback Impedance Control Architecture

Another way to decrease the inertia of the mechanical interface is to include positive motion feedback. To cancel the dynamics of the mechanical device, the compensation transfer function should be equal to the mechanical device impedance. This requires a good knowledge of the dynamic behavior of the interface. The major benefit of this strategy is that no force sensor is required; therefore, it is possible to obtain a simpler and cheaper final implementation. However, static friction cannot be compensated for because no change in force can occur without measuring a change in motion. Several algorithms based on this approach are proposed in [37, 62, 80].

5.9.4 Hybrid Compensation Impedance Control Architecture

The hybrid compensation strategy combines the benefits of the feed-forward and the positive feedback compensation methods. The best of both worlds can be attained by including forces yielded by the two methods in the same algorithm. Some of the algorithms that use this approach can be found in [37, 380].

5.9.5 Admittance Control Architecture

Admittance compensation includes a position controller that makes the system follow a trajectory imposed by a certain desired dynamic and computed in an admittance manner. This can be achieved by replacing the dynamics of the interface experienced by the user with the desired dynamics. The position controller is responsible for the movement of the device in unconstrained motion. This architecture has been used by Carignan and Cleary [62] and is referred to as "admittance control with position feedback." This concept is depicted in Fig. 5.19, where X represents the sampled position and F corresponds to the rendered Force.

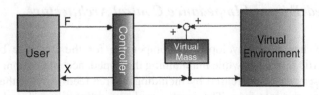

Fig. 5.19 Admittance control architecture

5.9.6 Position Feedback Admittance Control Architecture

In this architecture, the behavior of the virtual environment is introduced as admittance (also called the admittance display). If there is no virtual contact, the dynamics of the environment are replaced by desired dynamics in free movement. If, however, the controller gain is sufficiently large compared to the environmental impedance, the dynamics of the mechanical device is cancelled. This strategy has been used by different researchers, such as [10, 11, 228, 229, 381].

Some strengths and weaknesses of the various control algorithms are listed in Table 5.1. When comparing these approaches, several observations should be noted. On the one hand, the impedance-based approach with force feedback shows the most promise when stability can be maintained [62]. Furthermore, the impedance error is inversely proportional to the level of force feedback used, which in turn, is limited by the system's stability. On the other hand, admittance control offers more immunity to instabilities due to its filtering effect on the force signal. Additionally, the disturbance rejection properties are attractive. Therefore, it cannot be concluded that one control approach is clearly superior; the HAVE application and the characteristics of the haptic device determine the choice.

Table 5.1 A summary of control algorithms used with force feedback devices

Control algorithm	Summary	Advantages
Impedance interaction	Contact forces are computed using impedance causality and are restored via a negative feedback. Used with small inertia	Simplicity Accuracy
Impedance interaction with feed-forward compensation	Inertia is decreased by including force feed-forward to move the interface. Used with moderate inertia	Reduced device dynamics Accuracy
Impedance interaction with positive feedback compensation	Inertia is decreased using positive motion feedback. To ensure transparency, the compensation transfer function should be equal to the mechanical device impedance	Transparency Stability Inexpensive
Impedance interaction with hybrid compensation	Includes the feed-forward compensation and the positive feedback compensation forces	Stability Transparency Accuracy
Impedance interaction with admittance compensation	Includes a position controller that follows a trajectory imposed by a certain desired dynamic	Transparency Accuracy
Admittance interaction and compensation	The behavior of the virtual environment is introduced as admittance. Used for devices with high levels of nonlinear friction and high gear ratios	Transparency Stability Extendibility Disturbance rejection

5.10   Benchmarking Haptic Rendering Systems

Since the study of haptic rendering fidelity and realism is of significant interest, there has been increasing development of verification and validation algo-
rithms/techniques for haptic rendering systems. However, these efforts are impeded
by two main challenges. First, the vast majority of haptic systems are fundamentally
interactive, which prevents the consistent reproduction of results, and thus the
evaluation of a haptic system. Second, it is difficult to compare haptic data to real
data, as this requires applying identical forces to both real and virtual objects. In this
section, we present current efforts to evaluate the performance of haptic rendering
systems.

5.10.1   Existing Techniques for Haptic Rendering
System Evaluation

Initially, general-purpose systems for validating the physical fidelity of haptic
rendering systems were introduced. For instance, an impedance-based metric for
evaluating haptic devices was introduced in [80]. The authors discussed factors
affecting the dynamic range of haptic displays, such as the inherent damping,
the sensor resolution, the sampling rate, and the velocity signal filtration. Also,
Hayward and Astley defined a standard set of performance measures for evaluating
and comparing physical haptic devices [153]. The proposed measures include the
peak force, peak acceleration, and frequency-dependent measurements. In all, these
metrics capture basic performance characteristics of the hardware, but they only
partially characterize the performance of the complete HAVE system.

In fact, the performance of the haptic interaction requires accounting for both
the device properties and the rendering techniques, which are the hardware and the
software components of the haptic rendering system, respectively. To accommodate
such requirements, Kirkpatrick and Douglas presented a taxonomy of haptic
interactions and proposed the evaluation of complete haptic systems based on
distinct styles of haptic perception usage [209]. Other researchers, such as [117],
evaluated the effectiveness of specific haptic systems for particular motor training
tasks. For instance, the performance in [117] was measured in terms of position,
shape, timing, and drift. Meanwhile, Guerraz et al. [141] proposed a methodology to
evaluate a user's behavior and the suitability of a device for a particular task based on
physical parameters coming directly from the device itself. However, these metrics
are not general-purpose and do not address the realism of specific algorithms.

Raymaekers et al. [301] proposed an empirical method for evaluating haptic
rendering algorithms for correctness and performance. The evaluation procedure
includes the collision detection algorithms and force generation techniques. A number of reference objects (convex and concave) are explored for a certain period. The
recorded samples are played back using another algorithm, and the execution times,

the result of the collision step, the surface contact point, and the force vector are compared.

In order to evaluate the realism of interaction, Ruffaldi et al. [319] proposed an objective and deterministic procedure for haptic rendering algorithm verification and comparison. First, forces are collected from physical scans of real objects. Second, polygonal models of these objects are created. Third, different haptic rendering systems are used to interact with the models, and the forces are recorded. And finally, the haptic interaction results are compared to real-world interactions. The authors demonstrated the approach's ability to quantitatively assess haptic rendering systems.

5.10.2   A Framework for Haptic Rendering Systems Evaluation

This section presents an evaluation and comparison of haptic rendering systems. We have developed a framework that addresses the need for objective, deterministic, and possibly standard haptic system validation and verification. We conduct a performance evaluation and analysis of existing haptic rendering systems by changing the collision detection algorithm component. Several detection algorithms (for which implementations were available) are used to demonstrate the ability of the framework to evaluate the quality of the haptic rendering algorithm. The evaluated algorithms are the linear programming based I-COLLIDE algorithm, the polygonal-based V-COLLIDE algorithm, the DEEP algorithm, and the SWIFT++ algorithm (as presented later in Sect. 5.9.3).

The proposed framework strives to measure the correctness and efficiency of the rendering systems. First, the framework evaluates the rendering realism. This is done by computing the errors between forces generated from an actual physical interaction and those computed using a specific haptic rendering system under test conditions (comparing the haptic rendering algorithm output to reference "golden" data that was captured from real physical interactions with the modeled object) [7]. Second, this framework is capable of evaluating the haptic rendering system's ability to detect collisions. To measure the collision detection capabilities, we use two performance measures: the Collision Detection False Positives (CDFP) and the Collision Detection False Negatives (CDFN). The CDFP occurs when a collision is reported by the haptic rendering system and there was no actual collision. A CDFN is reported whenever a collision goes undetected. Finally, the framework enables performance evaluation of various haptic rendering systems by comparing, for instance, the time required for the processing of a predefined set of inputs. In short, it evaluates the speed of the haptic rendering system to compute the force feedback responses.

An overview of the performance evaluation pipeline is shown in Fig. 5.20. The evaluation procedure comprises two steps: a pre-processing stage followed by a processing stage. Since the haptic rendering system typically requires two sources of input, namely a geometric model of an object of interest and real-time positional

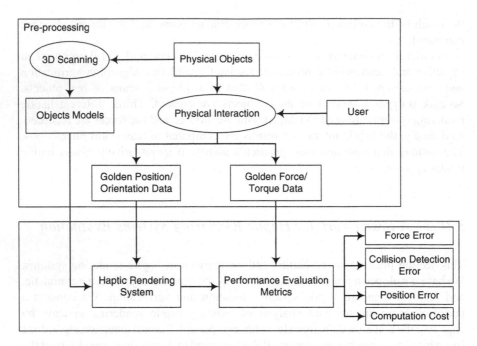

Fig. 5.20 Overview of the haptic rendering system evaluation pipeline

data collected from a haptic interface, such data must be prepared in the pre-processing stage. The pre-processing stage includes constructing a geometric model of a real-world object – typically using a 3D scanner – and collecting a series of correlated interaction forces and positions on the surface of that object with the user in the loop. The interaction data is then split into "golden" position/orientation data that will be fed to the haptic rendering system and "golden" force/torque data that will be used as reference patterns for the rendered forces/torques.

In the processing stage, the geometric model and the position/orientation data are input to the haptic rendering system. The output of a haptic rendering system is typically a stream of forces that must be sent to the haptic interface. A key goal of this analysis is to compare the rendered forces to the real-world data. The closer the rendered forces are to the real-world forces, the better the haptic rendering system. Furthermore, the performance evaluation includes determining position errors, computation costs, and false positive/negative collision detection rates.

5.10.3 Performance Comparison

Many software libraries have been developed in computer science to solve the problem of collision detection. In this section, we present a quantitative evaluation for a representative set of collision detection toolkits. The algorithms are representatives of convex polytopes and polygonal algorithms. Nonpolygonal algorithms are not considered in this analysis for two main reasons: they are the least used in haptic applications and converting the models from polygonal to nonpolygonal representation results in a declination in the model quality, which biases any performance evaluation.

In this section, we describe several algorithms used in testing and demonstrating the ability of the proposed framework for evaluating the quality of haptic rendering algorithms. The considered algorithms are: V-COLLIDE, I-COLLIDE, DEEP, and SWIFT++. We briefly summarize each algorithm, describe the experimental setup, and discuss the performance results.

1. DEEP [208]: This algorithm represents penetration depth-based algorithms. DEEP estimates the penetration depth and direction of the HIP whenever a collision is detected between two overlapping convex polytopes. The algorithm computes a locally optimal solution by walking on the surface of the Minkowski sum of the two polytopes. As described in [101], the penetration depth is defined using the features on the configuration space obstacle (CSO). DEEP is designed and implemented on top of the SWIFT++ and QHULL libraries [4]. Version 1.0 has been used in the comparison analysis.

2. V-COLLIDE [177]: V-COLLIDE is designed for large environments and can handle only triangular polygonal models. The algorithm reports when two objects collide by tracking the positions of the objects, thus a collision is reported to object precision. The algorithm uses temporal coherence between successive steps of a simulation to improve the performance of the dynamic environment simulation. Since the algorithm does not report penetration depth, we have implemented a simple force generation algorithm to estimate the force response for a collision; it is based on the force generation technique described in Sect. 5.4. The experimentation has been conducted using version 2.01 of V-COLLIDE.

3. I-COLLIDE [79]: I-COLLIDE is a linear programming-based collision detection algorithm. It is designed for interactive and exact collision detection in large-scale virtual environments, such as walkthrough scenarios. The algorithm reports the separation distance between objects by utilizing the Lin and Canny (LC) incremental distance computation algorithm [226] and an algorithm to check for collisions between multiple moving objects. I-COLLIDE models must be represented as convex polytopes, so a conversion from triangular mesh to convex polytope representation was necessary to experiment with this algorithm. The method computes the closest feature pairs – based on Voronoi regions – and calculates the Euclidean distance between the features to detect collisions. In our study, we have used version 1.3 of I-COLLIDE.

4. SWIFT++ [101]: This is a family of algorithms for the proximity query of closed and bounded polyhedral (polygonal) models. The library addresses the following queries: intersection detection, tolerance verification, exact minimum distance, approximate minimum distance, and disjoint contact determination. The library uses a decomposer to break down the boundary of each polyhedron into convex patches and creates a hierarchy of convex polytopes (convex hulls). Eventually,

pairs of convex hulls are tested using the LC algorithm. SWIFT++ version 1.1 was used in our evaluation.

As for the testing data, we used the same data set as in [7] where the authors developed a graphic model for a physical object (a duck) and recorded physical interaction forces with the real object (the golden data). The duck model, represented in the Waterfront .obj file format, and a contact trajectory were used to compare the relative force errors produced by the four algorithms. A laser range scanner was used to create the graphic model of the duck, which comprised 1,396 vertices, 1,456 normals, and 2,983 faces (in the case of I-COLLIDE, the model consisted of 83 polytopes, 36 faces each). Furthermore, optically tracked force sensors were used to acquire the haptic data (using the HAVEN facility at Rutgers University [7]). Comparing the rendered forces and the measured interaction forces (golden data) provides a quantitative basis for evaluating the haptic rendering system. To limit the comparison to collision detection techniques, we have used the same force computation and control algorithms for all four rendering algorithms.

We measured the performance of the four collision detection algorithms using the haptic rendering system evaluation pipeline, as shown in Fig. 5.20. Table 5.2 shows the performance metrics for the four algorithms. The total simulation time was 63 s. The performance metrics include the following measures:

1. Average run time: The runtime of an algorithm is the duration of one step of the simulation, which involves one iteration of collision detection without rendering time. The average run time was computed by taking the mean of all the trajectory simulation steps. All the times presented in this analysis were obtained on a Windows PC with 1 GHz Pentium III CPU and 256 MB memory.
2. Force error: is evaluated as the mean-square error (MSE) between the rendered and physical forces, computed as shown in (5.1).

$$\text{MSE} = \frac{\sum_{i=1}^{n} \sqrt{(F_{xr} - F_{xp})^2 + (F_{yr} - F_{yp})^2 + (F_{zr} - F_{zp})^2}}{n}, \qquad (5.1)$$

where (F_{xr}, F_{yr}, F_{zr}) are the force components of the force rendered using one of the algorithms and (F_{xp}, F_{yp}, F_{zp}) are the components of the real physical interaction force; n is the total number of samples in the simulation.

Table 5.2 Performance evaluation metrics for the four algorithms

Algorithm	Execution time (ms)		Force error (N)	CDFP (%)	CDFN (%)	Memory overhead (MB)
	Mean	Standard deviation				
DEEP	0.5788	0.0881	0.119604	0.720	0.0471	6,692
V-COLLIDE	0.4011	0.1511	0.153981	0.978	26.215	6,828
I-COLLIDE	2.8333	1.1515	0.434228	3.022	17.052	5.660
SWIFT ++	0.5057	0.1011	0.181107	0.864	0.0471	6,303

3. Collision detection false positives (CDFP): The CDFP occurs when a collision
   is reported by the haptic rendering system and there was no actual collision.
   It is calculated as the number of times a false collision is detected over the total
   number of collision tests during the simulation. In the "golden" data, the total
   number of collision tests was 31798 and the number of collisions was 4,856
   collisions.
4. Collision detection false negatives (CDFN): A CDFN is reported whenever
   a collision goes undetected. The CDFN is computed as the number of times a
   collision went undetected relative to the total number of collision tests.

We observed that the DEEP algorithm showed the best collision detection
accuracy (it had the minimum false positive/negative rates and minimum force
error); this is achieved at the cost of a relatively higher execution time. In Fig. 5.21,
we plot two lines: the bold line displays the rendered forces using the DEEP
algorithm, whereas the other line shows the physical interaction forces. We also
noticed that the DEEP algorithm showed a nearly constant execution time (the
standard deviation of execution time was 0.0881, as shown in Fig. 5.22). On the
other hand, V-COLLIDE had the least execution time at a cost of having the highest
memory overhead. Finally, since DEEP is built on top of SWIFT++, they both
showed similar performance with a slight difference in the force error. Unlike
SWIFT++, DEEP computes the penetration depth, and thus results in higher
accuracy force computation. The I-COLLIDE algorithm had the worst performance
because of the lack of coherence in the end-effector movement. As shown in
[79], this algorithm was designed for multi-body, large-scale environments where
it showed promising performance. Also, it showed better performance in terms
of memory overhead (see Table 5.2). Therefore, our conclusion is that collision
detection algorithms perform differently based on the application scenario that
defines the desired accuracy, speed, and/or model complexity, and the available
resources, such as computation power and memory allocation.

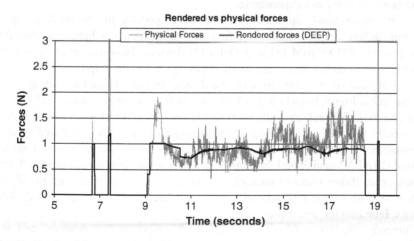

Fig. 5.21 Rendered forces vs. physical forces over time

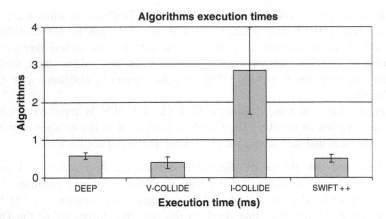

Fig. 5.22 The mean and standard deviation of the execution time

5.11   Haptic Software Frameworks

Early haptic/graphic libraries, such as Microsoft's DirectInput, were designed for a specific device (e.g., the SideWinder family of game controllers) to give access to input data by communicating directly with the hardware drivers and allowing different features for programing the mouse, keyboard, or haptic feedback joystick devices. Immersion's TouchSense API [186] was also originally developed for the specific Wingman force feedback mouse. In addition, the GHOST SDK, created specifically for SensAble Technologies' PHANToM device family, is still present; however, the OpenHaptic tool kit from the same company is a new, extensible architecture that offers additional capabilities [334]. The common factor in those libraries was the device-dependent approach in the software design, but more importantly, they are not expandable.

On the other hand, the Reachin API, developed by the Swedish company Reachin Technologies, became the first device-independent haptic/graphic API by supporting PHANToM [8] and Delta [1] devices. However, even though the Reachin API removes the concern of dependency on the haptic interface, it is still not expandable in terms of software design. In order to tackle such an issue, Novint technologies launched a set of graphic/haptic, open source APIs called e-Touch [267]. E-Touch also allows both PHANToM and Delta devices to expand and modify the API by allowing programmers to create a haptic-based desktop environment. Unfortunately, e-Touch is no longer available. However, a project entitled CHAI 3D envisioned the importance of a set of open source graphic/haptic libraries that allows users to interact with high or low-level code and modify the control algorithms for a variety of current and even future devices (PHANToM, Omega, Freedom 6S, and more).

Currently, the trend of haptic technology development is increasing, and as a result, software application interfaces are also evolving. As a consequence, the spectrum of haptic interfaces available as commercial or research and development products is growing. Based on the software process principles and lifecycle models, the APIs can be classified according to scope in terms of hardware applicability.

5.11.1 Commercial Application Program Interfaces

5.11.1.1 Microsoft DirectInput

DirectInput is an API that enables an application to retrieve data from input devices such as a mouse, a keyboard, and the force feedback "SideWinder" family of joysticks, and can be used with any brand of game controller [257]. Through action mapping, the API enables one to establish a connection between input actions and input devices that do not depend on the existence of particular device objects (such as specific buttons or axes). It supports keyboard properties and joystick slider data.

5.11.1.2 Immersion's TouchSense

Initially, Immersion's TouchSense API enabled the Wingman force-feedback mouse to deliver a rich array of sensations [186]. The API has since been extended to support touch-enabled devices, such as joysticks, steering wheels, and game-pads used in computer games, to transfer forces to a user's hand or fingers. The effects of period vibration, positional texture, enclosure and spring, directional constant, ramp, resistive damper, friction, and inertia can be supported.

5.11.1.3 Immersion's VHTK

VHTK stands for Virtual Hand Tool Kit and is the API for Immersion Corp's hand exoskeleton interaction devices [132]. VHTK was created for the exclusive use of Immersion's line of 3D interaction devices. This API supports only their multiple-contact haptic devices, which are integrated by three pieces of hardware: the CyberGlove, CyberForce, and CyberGrasp units. Multiple-contact haptic devices have multiple points of contact represented in the virtual environment, like a hand. The CyberGlove collects data that is related to the hand, such as the bending of the joints of each finger. The CyberGrasp is capable of generating a force in the medial and distal phalanx of the finger, with the exception of the pinky. The CyberForce armature has two functions: tracking the movement of the hand and generating force feedback that simulates inertia. The CyberForce armature allows for 6-DOF movement and is also capable of measuring hand rotation and translations. The Virtual Hand SDK/toolkit provides the following software capabilities:

- Offers an object-oriented model with an accompanying C++ library.
- Provides a general framework for constructing hand-enabled simulations from scratch or for integrating hand interactions into existing applications.
- Offers real-time collision detection capabilities between 3D digital objects.
- Provides a force feedback interface for CyberGrasp and CyberForce users.
- Offers full network support. A user can run an application on a host computer while getting device data from another machine, permitting interaction with geographically distributed teams.
- "Ghost-hand" support for managing position-tracker offsets to prevent the graphical hand from passing through objects.
- A fast polygon-level, collision-detection engine, including an open API for support of specialized third-party collision modules.
- An open API for model import and interfacing with third-party visualization software. A VRML/Cosmo (SGI Optimizer 1.2) implementation is included.
- Significantly improved overall structure with better run-time integrity and more complete error handling.
- A complete set of open source demonstration applications showing how each of the toolkit features can be used in your development.

In summary, the VHTK already meets the main requirement for decoupled haptics and graphics; however, the API is highly device dependent and not extensible.

5.11.1.4 Sensable's GHOST

The GHOST SDK is an abbreviation for General Haptics Open Software Toolkit. GHOST is an API that must be purchased from Sensable and is compatible with Sensable's Premium and Desktop PHANToM devices only [8]. GHOST offers a mid-level programming library, compared to other products from the same company, such as OpenHaptics. With a mid-level library, there is no need to deal with the low-level implementation for collision detection or force calculations. Therefore, this architecture eliminates a lot of the extra work that would have been required with a low-level library. GHOST is characterized by being quick and easy to learn, and many haptic functions are available for quick implementation through a well-documented library. Most importantly, GHOST's programming structure allows the user to entirely decouple the haptics pipeline from the graphics one, allowing the developer to manipulate the graphics library of their choice. Despite its attributes, GHOST is dedicated to support the PHANToM family, which makes it extremely device dependent. In addition, the API does not allow the extension of sophisticated control algorithms applied on the haptic rendering process.

The key features of the GHOST SDK include:

- The ability to model haptic environments using a hierarchical haptic scene graph.
- The ability to haptically render disparate geometric models within the same scene graph.
- The specification of surface properties (e.g., compliance and friction) of the geometric models.
- The use of behavioral nodes that can encapsulate either stereotypical behaviors or full, free body dynamics.
- General support for the generation of haptic human–computer interfaces, including effects such as springs, impulses, and vibrations.
- An event call-back mechanism to synchronize the haptics and graphics processes.
- The ability to automatically parse and use the static geometry of VRML 2.0 to generate haptic scene graphs.
- Extensibility through the subclassing metaphor.

5.11.1.5   Sensable's OpenHaptics

OpenHaptics is another API that Sensable provides with their line of haptic devices [334]. It is compatible with their entire line of PHANToM haptic devices. OpenHaptics is an API that allows both high- and low-level programming for haptics application development through the Haptic Library API (HLAPI) and the Haptic Device API (HDAPI), as shown in Fig. 5.23. High-level programming (through HLAPI) usually spares the user from a lot of the implementation logic, such as device control, collision detection, force calculation, while providing more control to the devices and haptic rendering algorithms. Sacrificing control is acceptable when one is only interested in rapid application development through use of the API. High-level programming tends to be easier and quicker because a lot of low-level programming implementation is not accessible to the programmers. The high-level library can be valuable in that it provides enough control for rapid prototyping.

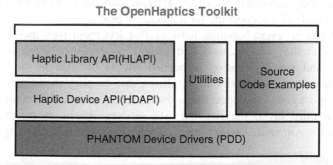

Fig. 5.23 The OpenHaptics API's components (adapted from Sensable Technologies Programming Manual [334])

The high-level library provides many desirable abilities, such as control of haptic virtual environments and special haptic effects. It allows users to quickly convert existing computer graphics applications into haptics applications. In the high-level library, the user is able to create a haptic object simply by describing the dimensions of the object. This is similar to the well-known OpenGL graphic library.

The low-level library (HDAPI) allows for graphics and haptics processes to be decoupled, thereby fulfilling an important requirement for more sophisticated prototyping. However, low-level programming inherently requires users to perform more implementation than higher level programming does. In order to create an application, the low-level library requires the user to develop the haptic implementation from the ground up. Using this library means that users develop their own algorithms for collision detection as well as their own force calculation algorithms. Therefore, HDAPI is more suitable to researchers who develop and test new haptic rendering and control algorithms.

5.11.1.6   Reachin API

The Reachin API allows the development of HAVE applications by using the C++ programming language and a combination of the Python script language and the descriptive markup language VRML (Virtual Reality Modeling Language) [385]. The idea behind this platform is a C++ API based on the VRML scene graph model. The Reachin API forms a hierarchical data structure from high-fidelity features as well as a complete set of classes, nodes, and interfaces for managing and synchronizing haptics, graphics, and audio in advanced 2D and 3D applications in a hierarchical data structure. Figure 5.24 shows the overview of the architecture of the Reachin API.

The basic mechanism that links all the scene graph components together is the event handling field network. The Reachin API model handles a rendering engine that uses one single scene graph manager that maintains the integrity of the scene (Multi-sensory Scene Graph Manager). Through the field network, the API defines the interaction between objects and other dynamic elements in the scene. Figure 5.25 shows an example of a scenegraph for a touchable box.

The API has an event handling network that lets them pass messages to one another so that they can respond and change. In addition, some sort of data dependency and synchronization is needed and is handled by this approach. The event handling capabilities layer allows one to plug in different modules in a coherent way and to define one's own interaction models within the scene graph. As a third layer, the Reachin API renders objects in pixels by interfacing with the OpenGL Library and renders forces through the haptic device driver.

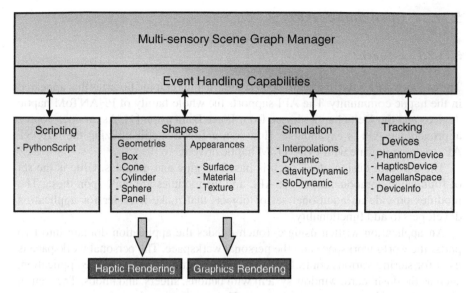

Fig. 5.24 A conceptual overview of the architecture developed in the Reachin API (Adapted from Reachin API) [6])

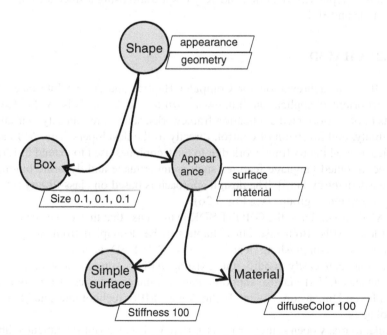

Fig. 5.25 The scene graph for a touchable box (adapted from Reachin Programming Manual [6])

5.11.2 Open Source APIs

5.11.2.1 e-Touch

The e-Touch API is the first set of open source, haptic/graphic libraries that appeared in the haptic community. The API supports the whole family of PHANToM haptic devices and the Delta haptic device from Force Dimension [1]. As an open source approach, the API is expandable. However, e-Touch is still utilizing the GHOST API to communicate with PHANToM haptic drivers.

The software design of e-Touch is based on glue and modules. Glue is the set of foundation components for the API, and the modules are built upon these. The modules provide an additional set of objects that make it easier for application developers to add functionality.

An application written using e-Touch divides the application domain into two parts: the world workspace and the personal workspace. The personal workspace is used for storing various controls and indicators that are specific to the application, such as the dashboard, window system with buttons, sliders, and knobs. The central object in an e-Touch system is the user. The user holds references to all the other objects that are vital for forces and the graphic rendering process. Unfortunately, e-Touch API popularity is fading, and very few are still using it since the emergence of CHAI 3D and H3D.

5.11.2.2 CHAI 3D

CHAI 3D is an abbreviation for Computer Haptics and Active Interfaces. It is an object-oriented application framework written in C++. This API takes the benefits of object-oriented application frameworks, such as modularity, reusability, extensibility, and inversion of control, directly to the developers. This API can be classified as a whitebox framework due to the techniques used to extend it. It relies on object-oriented language features such as inheritance and dynamic binding to achieve extensibility [115]. The design approach is based on class-library sets that are organized into 9 groups (see Fig. 5.26).

CHAI 3D is similar to the GHOST SDK in the sense that they can be categorized as mid-level APIs. Both also allow haptics to be decoupled from the graphics. However, one distinguishing feature is that CHAI 3D is a free, open source API, so one can easily modify the existing API to support new or prototype haptic devices. CHAI 3D also supports many commercial devices from companies such as Force Dimension, Novint Technologies, MPB Technologies, and Sensable Technologies.

Similar to many open source projects, CHAI 3D is not a well-documented library when compared with other commercial APIs such as the GHOST SDK. However, CHAI 3D is a good start for developing the guidelines needed to build a generic and extensible framework that allows the integration of software, hardware, and application design.

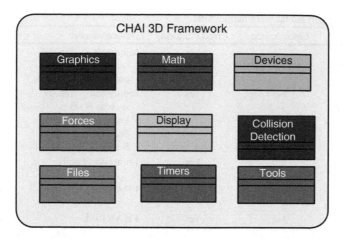

Fig. 5.26   Overall architecture of CHAI 3D [85]

5.11.2.3   H3D

H3D is a haptics application framework similar to the Reachin API, except that it is an open-source framework. It extends X3D, which is used instead of VRML to create a virtual environment. H3D also supports a variety of operating systems (Windows 2000/XP, Linux and Mac OS X). It is also based on the scene-graph approach, which is developed entirely in C++. It is dependent on OpenGL for graphics rendering and on HAPI (an open-source, cross-platform, haptics rendering engine [2]) for haptic rendering. HAPI adopts various haptic rendering algorithms and supports most commercial haptic devices in a similar way to CHAI 3D. In addition, H3D is built using many software industry standards, including STL, XML, and X3D.

5.11.2.4   The MOTIV Haptic Development Platform

The MOTIV Haptic Development Platform is an SDK that offers developers the ability to add tactile feedback into their mobile device applications (in particular, the Google Android operating system) [185]. The SDK is composed of a library of pre-designed haptic effects that are directly accessible through the API, which eliminates the complexities of integrating haptic modality into pervasive devices. The API works in conjunction with Immersion's TouchSense™ technology [186]. MOTIV has three distinguishing features: (1) a UI module that integrates haptics into the Android OS user interface, (2) a theme manager module that provides a list of haptic themes to further customize the customer's device, and (3) a reverb module that automatically translates audio data into haptic effects.

Table 5.3 presents a comparison summary of haptic APIs, where the "Device support" field represents the device families (manufacturers) the API supports.

Table 5.3 Comparison between haptic APIs

API	Open source	Cross platform	Device support	Language
Microsoft DirectInput	No	No	Microsoft family such as Xbox 360 controllers	C
Immersion's TouchSense	No	Yes	Immersion family such as 6000 Series, 6100 Series, and 6500 Series	C++
Immersion's VHTK	No	Yes	Immersion family	C++
Sensable's GHOST	No	No	PHANToM	C++
Sensable's OpenHaptics	No	Yes	PHANToM	C++
Reachin API	No	No	PHANToM, Delta, Falcon	VRML/C++/ Python
e-Touch	Yes	No	PHANToM, Delta	C++
CHAI 3D	Yes	Yes	PHANToM, Delta, Falcon	C++
H3D	Yes	Yes	PHANToM, Delta, Falcon	C++
The MOTIV	Yes	Yes	Android phone	C++

5.12   Closing Remarks

Computer haptics deals with (a) the design and development of algorithms and software APIs to model haptic physical properties for virtual objects and (b) computing the interaction forces between the haptic interfaces and the manipulated HAVE environment. One of the main research and development issues in computer haptics is dealing with the delivery of collision detection and force response stimuli. Force response calculation is related to the emulation of the end effector, usually the haptic interface point (HIP), and the penetration depth simulation when interacting with other objects and/or the physical (haptic) properties associated with the object(s). In general, researchers have investigated approaches to obtain promising performance results for the haptic rendering algorithms. These results depend on the number of DOFs of the haptic display and the sophistication of the haptic models, which are tightly tied to the used physical formulation and the collision detection algorithm.

Collision detection and rendering algorithms have been comprehensively discussed in the field of computer graphics. In the context of haptic applications, those algorithms cannot be directly applied due to their slower update rate when compared to the moderate update rate requirement of 1 kHz for stable haptic interaction. Graphic rendering algorithms are based on finding collisions among all objects populating the virtual world. This is unlike haptic rendering, where collisions occur in the vicinity of the haptic interaction point (between the haptic interaction point and the virtual world). This is why haptic rendering algorithms use localized collision computation around the HIP through hierarchical bounding volumes.

In the early stages, haptic application program interfaces were dominated by commercial products, such as Immersion APIs, Sensable's GHOST and OpenHaptics libraries, and the Reachin API. Later, the need for development of nonproprietary software frameworks motivated some open-source, haptic-based libraries, including CHAI 3D, H3D, MOTIV API, and e-Touch. Finally, we are currently witnessing an increased interest in HAVE application development, especially in the gaming industry.

Chapter 6
Multimedia Haptics

6.1 Introduction

A variety of haptic interfaces and rendering methods have proliferated HAVE application development and enabled more immersive and interactive experiences with
virtual and mixed environments. For example, force feedback-enabled surgical
training systems, such as *da Vinci*,[1] and cheap force feedback devices, such as the
Novint Falcon [2] (which gives video game players more vivid gunshot feelings in
shooting games), are available on the market. However, haptic-enabled applications
have not yet been widely used nor are they considered easily accessible. A potential
reason for this is the lack of more general-purpose content and its dissemination.
Multimedia haptics research targets this problem by incorporating haptic modality
into multimedia systems in order to more easily create, store, and deliver haptic-
enabled applications and content. This chapter broadly covers haptic content
creation, representation, transmission, and standardization.

As an analog of audiovisual media, haptic media presents haptic properties and
data that will result in touch stimuli. It contains the contents that are displayed
through touch as well as the information as to how these contents are arranged in
time when they need to be played back. For example, digital video basically consists
of sequences of static pictures and timing information for determining intervals
between these pictures. Although audiovisual media formats are wellestablished and
deal well with synchronization issues with other media and over network links, the
haptic media is emerging as a new media and has not yet been well established as a
widely usable format.

---

[1] www.davincisurgery.com.
[2] www.novint.com.

A. El Saddik et al., *Haptics Technologies*, Springer Series on Touch and Haptic Systems,   145
DOI 10.1007/978-3-642-22658-8_6, © Springer-Verlag Berlin Heidelberg 2011

6.2   Haptics as a New Media

Haptic media can be divided into two distinctive categories based on whether it is arranged in time or not; that is, whether the intermedia synchronization of haptic data is linear or nonlinear. Linear haptic media refers to haptic sensations that progress sequentially in time without any navigational control. This includes recorded movements or cutaneous patterns of touches to the skin, which create experiences of passive haptic playback. Nonlinear haptic media, on the other hand, is spatially arranged and offers users rich interactivity. This allows them to touch and explore a haptically displayed object, experiencing a compelling, active haptic interaction through both force and tactile information.

Essentially, linear haptic media encompasses skillful movements or forces and events, such as alarms, direction cues, textures, and on-screen motions (like the bouncing of a ball). One of the most used data types is motion data, which records positions, velocities, or forces of human body parts. In motor skill transfer applications, a skillful gesture, such as calligraphy, is captured through force feedback devices that are specifically designed for the gesture. In the calligraphy example, the position of the tool-tip corresponds to a pen tip and is sequentially recorded. The recorded data can then guide a user to follow the recorded path by applying position control through the haptic interface. Examples include: handwriting [242, 388], surgical skills [26, 76], and palpatory diagnosis [175].

Linear haptic media can be represented by a sequential series of actuation intensities, which can be deployed to control a grid of tactile stimulators spread over an area of skin. The intensity of these touch sensations may correspond directly to audio and video events in the content. This kind of tactile playback has been applied in movie and entertainment industries. For example, tactile stimulation was introduced with *Percepto*, which was used for the 1959 movie *The Tingler*, feedback was provided through vibrating devices attached to the theater seats. Recently, this type of haptic media has been widely used due to the development of tactile devices that can be used for instant messaging [318, 346], movies [206], music [142], letter displaying [400], directional cues in a car [364], etc.

Nonlinear haptic media can take the form of not only force and shape information but also the surface properties of an object's texture (friction, roughness, stiffness). In this case, viewers navigate in a haptic scene and feel objects only by exerting their own agency; they must actively explore the environment to feel the haptic cues. Nonlinear haptic media must also encompass object dynamics, i.e., how they move and behave in response to user input. This can include general terms, such as mass and inertia, but also more specific ones, such as spring constants defining the travel distance and sponginess of a virtual button. This kind of representation is arguably best displayed on commercially available force feedback devices. Nonlinear haptic media is the media type most used in haptic applications adopting virtual environments.

6.3 HAVE Content Creation

The adoption of haptic interfaces in human–computer interaction paradigms has led to the demand for new tools and systems that enable novice users to author, edit, and share haptic applications. While there are plenty of standard tools for capturing and creating audiovisual media in the market (due to its prevalent consumption), equivalent tools for haptic media are not widely available, and HAVE application development remains a time-consuming experience that requires programming expertise. Additionally, assigning physical material properties, such as stiffness, static friction, and dynamic friction, is a tedious and nonintuitive task because it requires the developers to possess technical knowledge about haptic rendering and interfaces. There is also a lack of application portability, as haptic applications are tightly coupled to specific devices, which necessitates the use of specific corresponding APIs. In view of these considerations, there is a clear need for an authoring tool that can build haptovisual applications while hiding programming details (API, device, virtual models) from the application modeler.

In this section, tools and methods that can accelerate haptic content generation are introduced. In a broad view, just as audio and video media can be directly captured from an environment through the use of cameras and microphones, or virtually synthesized through music synthesizers, 3D modeling, and animation tools, haptic media can be captured and synthesized in a similar way.

First, haptic media can be recorded using physical sensors. There are a few studies on the automatic capture of haptic surface properties such as stiffness, friction, and roughness [214]. In addition, the dynamic properties of haptic buttons have been acquired by measuring and analyzing the force profiles of real physical buttons [207]. Movement data can, for instance, be measured with a 3D robotic arm equipped with force-torque sensors or with a motion sensor, such as an accelerometer. As an example, in order to get more involved in a virtual soccer game, we can measure the kicking and bouncing collisions of the soccer ball by equipping it with internal piezo-electric impact sensors. These measurements can then be displayed to a haptic device [271].

Second, haptic media can be synthesized using specialized modeling tools. Tools for generating audiovisual media in the form of 3D modeling environments are commonplace, but there have been a few efforts to create modeling tools that integrate haptic properties into a 3D scene as well. Most of the previous haptic applications and contents are developed and generated using haptic SDKs and toolboxes such as OpenHaptics Toolkit, CHAI 3D, Handshake proSENSE Toolbox, etc., but it requires significant programming knowledge and skills. SensAble introduced the Claytools [387] and FreeForm systems [154] to incorporate haptics in the process of creating and modifying 3D objects. In these systems, the user receives physical feedback so that it feels as though they are physically sculpting the objects, but they still do not have the capability to apply haptic properties to the 3D objects. Reachin Technologies [217] introduced Reachin API as an object-oriented development platform that allows users to design haptic scenes by editing VRML-based script without programming, but it is not as intuitive as directly manipulating the scene by touch.

Fig. 6.1 A snapshot of the tactile authoring tool [206]

As an example of linear haptic media authoring, Gaw et al. [125] have developed a software tool that plays video and simultaneously shows and records a user's position (with a 3-DOF point device). This enables the tracing of movements in scenes that involve dynamic human motion (such as orchestral conducting). This spatiotemporal path can later be played back on a force feedback device, effectively providing a trace of the user's original movements in synch with the audiovisual content. In the tool by Kim et al. [206], tactile video can also be recorded through gestures made on touch sensors, such as touch pads or touch screens, as shown in Fig. 6.1. Essentially, as a video is played, a user can input patterns of tactile sensation by making movements on the surface of such a sensor. This creates a sequential stream of 2D information, which is synchronized with the audiovisual content and can be used to represent patterns of tactile activation.

On the other hand, there exist several nonlinear haptic authoring tools, such as K-HapticModeler [336] and HAMLAT [103], providing interfaces that support the construction of a 3D scene. They allow both haptic surface properties and dynamic movement properties to be assigned to parts of that scene. HAMLAT is based on

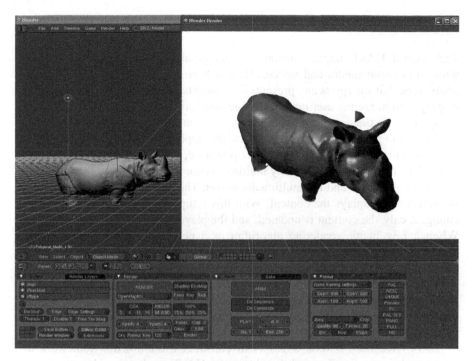

Fig. 6.2 A snapshot of the Blender-based HAMLAT editor with the haptic renderer [103]

the Blender[3] software suite. It is an open-source 3D modeling package with a rich feature set for creating and editing 3D objects. In HAMLAT, a 3D model is graphically designed, and haptic properties are assigned to the model through one of the input panels, as shown in Fig. 6.2. Users can check how the assigned haptic properties feel through a rendering model.

Finally, haptic media can also be derived automatically from analysis of other associated media. Consider first that music visualization (generating animated imagery based on a piece of recorded music) is an example of automatically converting audio media into visual media [351]. The same can be done for the automatic generation of haptic media from other media sources. While some associations are relatively obvious, others are potentially more rewarding. For example, the trajectory of a soccer ball or the forces exerted on a race car as it corners could be automatically extracted from video or animations using image processing techniques [399].

---

[3]http://www.blender.org.

6.4   HAVE Content Representation

The created HAVE haptic contents are temporally and spatially synchronized with audiovisual media and stored. They will be consumed through a software application that interprets and presents the contents through audiovisual and haptic displays and measures user input to enable interaction. Traditionally, there was not an apparent distinction between the haptic content and the software application, and they were packaged together. When the application behavior needed to be changed, either a software update was produced, or the application was totally re-implemented. However, in many multimedia scenarios there is a clear distinction between the content and the multimedia player. The two are independent; a player interprets and displays the content. With this setup, when the content scenario is changed, only the content is updated, and the player does not have to be updated. When a new haptic rendering algorithm or a new haptic device is developed, only the player or device driver is updated with the new algorithm. In order to separate haptic content from players, a haptic content framework is necessary to deal with the content creation, representation, delivery, and consumption. The content representation method is the most important part because it dictates how the haptic content is created, synchronized, and stored. The content is also systematically interpreted and presented following the representation method.

There have been several attempts to construct a multimedia framework that supports the easy creation and distribution of HAVE applications based on existing audiovisual content frameworks. The haptic broadcasting framework by Cha et al. [69] is based on the MPEG-4 framework and is intended to target broadcast applications. The MPEG-4 framework uses Binary Format for Scene, which inherits VRML. The representation method in this framework is very similar to the Reachin API and H3D by means of using a scene graph. It deals with linear haptic media too.

In contrast to the simple download-and-play delivery system found in VRML and X3D frameworks, the MPEG-4 framework supports streaming media, so haptic media can be streamed and consumed while being downloaded. For example, in [69] a tactile video is defined to represent a grid of intensities of tactile stimulation that can be mapped to an area of the user's skin and rendered through devices such as tactile arrays composed of a grid of actuation elements. As shown in Fig. 6.3, four frames of 10×4 pixels are illustrated on a timeline. Each pixel corresponds to an actuator on the glove-type vibrotactile device. In each frame, the whiter the color is, the more intense the actuation magnitude is. As a result, a user wearing the glove-type tactile device could feel a stimulus moving from the finger tip toward the wrist, covering more area and getting stronger over time.

It is very difficult to model a real moving scene with polygons or voxels, so new techniques are being developed. Consider, Fig. 6.4, which shows a dynamically changing 3D scene represented using color images and corresponding gray-scale depth images containing per-pixel depth information. The gray level of each pixel in the depth image indicates the distance from the camera. The higher (whiter) the level is, the closer it is to the camera. Since this depth image-based representation

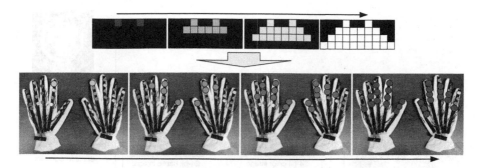

Fig. 6.3 A tactile video that corresponds to a set of actuators in a tactile device [69]

Fig. 6.4 A representation of a 2.5D video that has touchable geometry information and haptic surface

uses images for modeling a scene, a natural video that captures a real moving scene can be easily generated using stereo matching algorithms or a depth camera, such as the ZCam (see Fig. 6.4). In the research by Cha et al. [68], haptic surface properties, such as stiffness, static friction, dynamic friction, and roughness, are represented with images and mapped onto the 3D geometry defined by the depth images. In order to keep consistency with existing visual media, haptic properties are represented by 8-bit channels, composing a 24-bit color image, and roughness is represented in an alpha channel in the form of a height map image. Figure 6.5 shows a sample 3D scene and its geometric and haptic components.

6.4.1 Haptic Applications Meta Language

The Haptic Applications Meta Language (HAML) is designed to provide a technology-neutral description of haptic models. It describes the graphics of the environment (including the geometry and scene descriptions), the haptic rendering, haptic devices (the hardware requirements), and application information. In other words, HAML is the meta-language by which haptic application components,

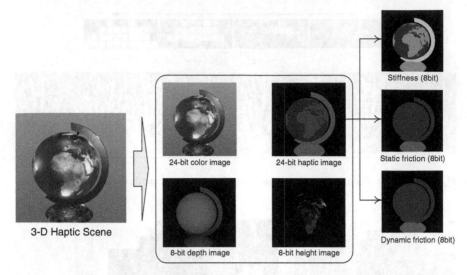

Fig. 6.5 A representation of a 2.5D video that has geometry information to be touched and haptic surface property images

such as haptic devices, haptic APIs, and graphic models, make themselves and their capabilities known. There have been at least three foreseeable approaches to implementing and utilizing HAML instance documents:

1. Application descriptions that define description schemes for various haptic application components. These can be reused in the future to compose similar applications, given equivalent requirements/specifications.
2. Feature descriptions where the HAML description is obtained from the device/API/model through a manual, semiautomatic, or automatic extraction and saved in a storage system for later use.
3. HAVE application authoring and/or composition.

   The basic components of the HAML framework are shown in Fig. 6.6. The user interacts with the HAML framework via the GUI component that captures the basic user requirements (the interaction type/device, the virtual environment components, data recording, etc.). These requirements are then passed through the translation engine, which relies on the HAML schema to "pump-out" a HAML-formatted document. This document contains a startup/default configuration of the haptic application, which is required for the framework to work. The Authoring Agent (AA) parses the HAML file and dynamically creates the haptic application by selecting and composing components – haptic device, rendering engines, collision detection engines, graphic components, and APIs – that meet the specifications defined in the HAML file. Notice that the HAML repository stores HAML-formatted descriptions for all available devices, all haptic and graphic APIs, and all related information.

Fig. 6.6 HAML framework

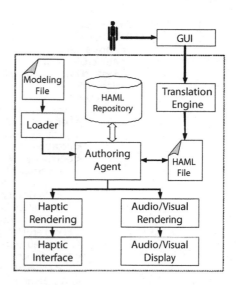

Technically, HAML is an XML-based schema meant to describe HAVE applications. The HAML schema is instantiated for compatibility with the MPEG-7 standard through the use of description schemes (DSs). The HAML structure is divided into seven description schemes: application, haptic device, haptic API, haptic rendering, graphic rendering, quality of experience, and haptic data descriptions. More details about HAML can be found in [102]. An excerpt of an HAML document is shown in Fig. 6.7.

Figure 6.7 demonstrates four DSs of the HAML structure: the application, the author, the system, and the scene. The application and author DSs organize high-level and general considerations of the haptic application, such as the application name, type, and author's name and contact information. The system DS describes the computer specifications (processor, operating system, network card, etc.) and the device/SDK/API support. Finally, the scene DS provides an object-based description of the graphic and haptic scenes. The scene comprises one or more objects, each identified by its type, location, orientation, geometry (model vertices and mesh topology), appearance, and haptic properties (stiffness, damping, friction).

6.5   Haptic Media Transmission

Most nonlinear haptic media data is small in size. Stiffness and friction can be represented by scalar values. Roughness can be represented by parameterized scalar values or a gray scale image. Dynamic parameters for describing haptic widgets, such as buttons, are scalar values. Since these values are static and spread over surfaces, their data representations in a whole virtual environment are significantly

```
<? xml version="1.0"    ?>
< HAML >
   < ApplicationDS   > ... </ ApplicationDS   >
   < AuthorDS   > ... </ AuthorDS   >
   < SystemDS   > ... </ SystemDS   >
   < SceneDS  >
      < Object  >
         < Type  > ... </ Type  >
         < Name  > ... </ Name  >
         < Location  > ... </ Location  >
         < Rotation  > ... </ Rotation  >
         < Geometry  >
            < VertexList  >
                  < Vertex  > 0, 1, 0   </ Vertex  > ...
            </ VertexList  >
            < FaceList  >
                  < Face  > 1, 2, 3  </ Face  > ...
            </ FaceList  >
         </ Geometry  >
         < Appearance  >
            < Material  > ... </ Material  >
         </ Appearance  >
         < Tactile  >
            < Stiffness  > 0.8 </ Stiffness  >
            < Damping  > 0.9 </ Damping  >
            < SFriction  > 0.5 </ SFriction  >
            < DFriction  > 0.3 </ DFriction  >
         </ Tactile  >
      </ Object  >
   </ SceneDS  >
</ HAML  >
```

Fig. 6.7 An excerpt from an HAML document

smaller than the geometric data and audiovisual media. However, nonlinear haptic media in tele-collaborative haptic environments pose new challenges at both the application and communication (networking) levels. Haptic interaction requires simultaneous interactive input and output through the haptic device with an extremely high update rate (up to 1 kHz). At the application level, improvements to consistency assurance, access control, transparency, and stability are undergoing extensive research. At the networking level, key quality of service parameters, such as network latency, jitter, packet loss, scalability, and compression, have been investigated and researched. This section fills the gap of understanding the characteristics of haptic interaction over a network and when multiple users are simultaneously interacting with the same environment. In the following subsections, the research about tele-collaborative haptic environments will be broadly introduced.

6.5.1   Classification

In the last few years, research has moved to include more implementations of shared virtual environments with the inclusion of haptics. Thus, applications on the implementation of these scenarios have been extended from training and education to gaming. This has led to a heterogeneous taxonomy or terminology for the different architectural styles of haptic applications. One classification of these interaction modes is presented in [95]. The authors identify three classes of shared environments: (1) static environments where the user can browse a stationary environment by feeling the haptic information stored in a document, website, database, etc., (2) collaborative environments in which users alternate in manipulating a shared environment, and (3) cooperative environments where users can simultaneously interact with the same object and feel their mutual force feedbacks.

There have been several taxonomies to classify haptic-based applications. For instance, the authors in [119] differentiated two types of interactions: independent and dependent. Independent interaction is the user interaction of multiple users within the same virtual scene that does not produce forces between users. Unlike independent interaction, and similar to a standalone interaction in terms of touch, dependent interaction exists when considering two or more users' interactions. New types of dependent interaction appear when considering two users' interactions (extendable to more than two users): user–user (mutual touch), user-object-user (cooperative task), and user-object-object-user.

Another classification that is based on the energy exchanged between users is presented in [314]. The authors classify haptic environments as either unilateral or bilateral. In the unilateral interaction, an operator is interacting by means of a haptic device, and the interaction data is sent to different sites to reproduce the operator's actions so that other users will feel the operator's manipulation. However, the operator does not receive haptic feedback from the other users. In bilateral interaction, all users can feel the other users' interactions. This obviously increases the sense of co-presence between users. In general, the haptic sensation felt by a remote user is indirectly computed and perceived through virtual scene modification.

In this context, many terms have been used to refer to multi-user virtual environments, including Collaborative Haptics [268], Shared Haptic World [156], Collaborative Haptic Virtual Environments [181], Cooperative Haptics [58], Distributed Haptic Virtual Environments [353], and Collaborative Haptic Audio Visual Environments (C-HAVE) [341]. When the haptic application disseminates over a network, the environment is addressed using several terms, such as Distributed Haptics [176], Tele-Haptics [342], and Networked Haptics [357]. Here, we adopt two classes to describe the works in collaborative haptic environments in order to eliminate any ambiguities due to terminological conflicts.

We will use the term Collaborative Haptic Audio Visual Environment (C-HAVE). However, under this class, the transmission of haptic information over a network, either dedicated or nondedicated, will be referred to as the Networked Haptic

Environments (NHE) subclass. The term will reflect the networking aspects of haptic data communication without considering collaborative scenarios.

6.5.2   Collaborative Haptic Audiovisual Environments

C-HAVE can be either networked or nonnetworked (standalone). Standalone C-HAVEs involve at least two entities interacting with a shared environment via two haptic interfaces. The environment comprises real, virtual, or a combination of both types of objects (augmented reality). In a standalone environment, both the environment objects and the state of the environment are stored locally. The fundamental concept here is to develop a realistic haptic/graphic model where tasks require multiple points of interaction, such as when grasping an object. The human–computer device interaction can be managed either in a synchronous or asynchronous fashion. The main issue is the synchronization and coordination between the haptic and graphical rendering loops.

The participants in the collaborative environment can be either passive or active. Active participants can apply manipulation forces and feel and observe the effects of actions by other users in the environment. Passive users, on the other hand, are viewers who can see and feel changes of the states of the environment but cannot alter its objects. In some cases, passive users' participation is limited to only acoustic and/or visual feedback.

The medical field is one example of an application field for such types of environments. In this context, there is the Virtual Haptic Back (VHB) project developed between two Ohio University departments: Engineering and Osteopathic Medicine [175]. It is a series of computer-based haptic simulations of the human body that assist students in the learning of palpatory techniques. The VHB system requires dual PHANToM 3.0 haptic interfaces, one for each thumb.

Figure 6.8 does not reflect the physical distribution of the nodes or the databases. For instance, the virtual objects database can be distributed over the participant host machines (objects that are owned by a node can be stored locally in that node). The diagram presents the logical architecture of the collaborative haptic environment. Notice that the two databases are dynamic because new participants and/or objects can join or leave the environment at run-time. The participant's database stores the identification information of the nodes (such as logical identifiers and IP addresses) that are currently interacting with the environment. Furthermore, the participant's database contains the QoS parameters associated with the links between all the connected nodes. The virtual objects database contains all the information about the objects that are populating the environment. This information includes the object identifier, owner identifier, static properties of the objects (shape, color, size, etc.), and dynamic properties such as position, orientation, and velocity.

The concept of sending haptic information over a network is referred to as "tele-haptics". Tele-haptics occurs when two users (for examples user A and user B), who are located at remote locations, use haptic interfaces to connect over a communications network. Each user is presented the same shared environment where he/she sees the environment and the two participants' avatars. For instance, whenever one of the users moves his/her haptic stylus, the avatar of that user changes its position in the local host. This new position information is sent to the other party (user B) to update the position of the avatar on the remote host. Therefore, the consistency between what each user sees will be assured. The same thing occurs when a haptic device avatar collides with an object in the environment; the new environment state will be sent to the remote host to resynchronize the two instances of the environment.

6.5.2.1 Issues in C-HAVEs

Most of the work performed has been focused on collaboration over an LAN or the Internet using a best-effort channel. In information technologies, there is an unavoidable time delay. LAN networks also experience this delay, although such a delay can be considered negligible (orders of microseconds). Moreover, it is not a fixed delay time. Every time a packet is sent, it can be received earlier or later than the previously sent packet. The variation of the network delay is known as jitter. Related to the fact that several streams of data may be circulating over the network, users may be exposed to lost or out-of-order packets. Another factor to be taken into account is the available bandwidth and throughput to deliver information. When designing C-HAVE applications, the following points should be considered as guidelines:

> *Stability.* This is always a key design consideration in any haptic system, and C-HAVE applications are no exception. Unwanted vibrations and unbounded forces are not only distracting but are also potentially unsafe for the human

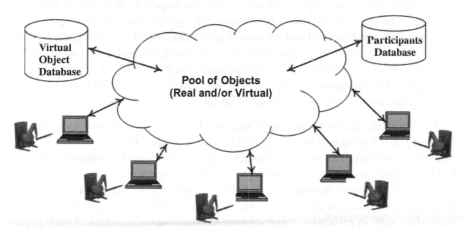

Fig. 6.8 A generic architecture of a Collaborative Haptic Audiovisual Environment

operator, especially in medical applications [291]. Many network impairments, such as delay, jitter, and packet loss, can easily result in unstable performance.

Fidelity. Haptic fidelity represents the quality of haptic sensations over a network. Theoretically, 100% application fidelity means that the user feels the remote user/environment as if they are local and touched directly by a user [401].

Heterogeneity. It is quite reasonable to assume that a typical networked C-HAVE application contains a pool of heterogeneous haptic devices with which users can interact within the shared environment. Modeling and interacting with a haptic device should depend upon many of the device's attributes, such as the number of degrees of freedom, the minimum and maximum forces, the workspace, and so on. Therefore, designing an abstract device-independent environment is the subject of recent research.

Scalability: Scalability measures the ability of a system to support a large number of participants. The issue of scalability in C-HAVE applications is different from that in virtual environments. In a typical virtual environment scenario, thousands of users participate in a shared environment and can see and hear each other and the environment (for example, in a virtual game environment). On the other hand, scalability of C-HAVE applications is limited to just a few users. We argue that the maximum number of collaborators in C-HAVE applications is ten active participants. This is similar to what we encounter in the real world where it is unlikely that more than ten users will co-touch the same object simultaneously. For example, in a C-HAVE Tele-surgery application there is a limited number of surgeons engaged in co-touching the patient.

Consistency. The concept of consistency implies that all the participants are experiencing the same exact state of the environment; what they view, hear, and feel is synchronized across all the participants. The consistency should be maintained at two levels: the graphic level and the haptic level. At the graphic level, the graphics of the distributed haptic application should have the same state for all the participants at any given time. At the haptic level, each participant should feel the same haptic interaction forces depending on the interaction paradigm. Furthermore, the consistency between the haptic, audio, and video scenes should also be maintained. Finally, consistency means that the application maintains a coherent state and resolves any conflicts arising from multiple users co-touching the same object.

Implementing a consistency assurance mechanism is an essential task when it comes to developing any peer-to-peer distributed virtual environment application. Such a mechanism is responsible for ensuring that every node participating in the C-HAVE application keeps a consistent view of the environment.

When it comes to consistency, distributed applications can be organized into two categories: discrete and continuous applications. Discrete applications change their state only in response to user-generated operations. Examples of such applications include distributed white boards and shared drawing tools [1]. Continuous applications, on the other hand, not only change their state in response to user-generated operations, but also to the passage of time. Examples of these applications include multiplayer games and distributed virtual environments in general.

The most popular consistency assurance mechanism used for such applications is Dead Reckoning (DR) (especially in gaming). In Dead Reckoning, it is assumed that the behavior of each object over time is predefined. The object state is locally modified by combining the received updates from networked participants to maintain consistency of the environment. The resulting state is calculated with respect to the one that could have been obtained from the DR algorithm. If the difference between these two states is larger than a preset threshold, an update is broadcast to all participants. The main difficulty with this algorithm is its potential to produce short-term inconsistencies during the transmission of an update destined to correct the error in the predicted state. To solve the problem of inconsistency, Mauve has proposed an alternative approach that uses the local lag and the timewrap algorithms [389].

Consistency mainly depends on the following factors: user actions, the virtual scene, and network conditions. First, a virtual scene can have static and/or dynamic objects (i.e., objects that can be virtually manipulated). Second, the user interaction within a virtual scene can include touching a static object, pushing a dynamic object, or grasping an object.

Local Lag Algorithm

In the local lag algorithm, it is assumed that each operation is qualified with two timestamps: the time when the operation is issued and the time when the operation should be executed at all nodes concurrently. The difference between these two timestamps is called the lag value. Figure 6.9 summarizes this concept. As it can be clearly deduced, the longer the local lag value, the less responsive the system will seem, however, using this algorithm, the local copy of the simulation on each node will be more consistent.

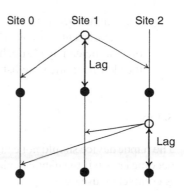

Fig. 6.9 The local lag concept

Choosing an appropriate lag value is important in order to balance the responsiveness of the system and its consistency. Mauve proposed the use of a constant local lag throughout the execution of the application [389]. The local value is extracted from two values: the maximum of the average network delays among all participants and the maximum allowable response time of the application. The latter value is obtained from psychological tests and is more or less subjective [278]. Chen has also proposed an adaptive approach for choosing the local lag value [74]. Using statistical sampling, the next minimum duration of network latency is estimated, and the local lag value is corrected accordingly.

Timewrap

Local lag by itself cannot solve the entire consistency problem. There will always be times where, due to jitter, packet transmission takes longer than the local lag time. In this case, these packets are processed locally before being processed by the other participants, resulting in inconsistencies. Mauve proposed the timewrap algorithm in order to deal with such short-term inconsistencies, thereby ensuring complete consistency for all the participants [18]. The timewrap algorithm involves periodically correcting any short-term inconsistencies by rolling back each loop to a base state where all the operations are received or locally produced, and the base state occurrence is re-executed. This can be summarized in the following steps:

- Receive an operation or produce an operation locally.
- Save the received or produced operation in the operation list in ascending order according to its t^*, where t^* indicates the time the operation should be executed.
- Save the state of the system, St, in the state list according to t, where t is the time when St was the state of the system.
- Every T milliseconds, where T is a preset period of time, perform the following:

  - Retrieve the base state Si, where Si was the state of the system T milliseconds ago.
  - Perform all the operations stored in the operations list between the time of Si and the current time by applying them to Si according to their chronological order indicated by t^*. Each time an operation is applied, the new state replaces the old one in the state list.

Safety

The haptic device should not cause any harm to the user. A C-HAVE application can become unsafe if the haptic device becomes unstable. In such a case, the application is completely useless!

Network Latency

When transmitting over a nondedicated network such as the Internet, data packets might be delayed by buffering, processing, transmission, or propagation. This delay varies depending on the physical distance between the communicating parties and in accordance to the state of the communication network.

Fukuda et al. [121] aim to solve the problem of distant haptic interaction. Their system is based on the idea discussed in [256] in which force feedback degraded when the delay was about 30 ms. Abrupt forces occurred due to the fact that the force is computed in proportion to the interpenetration between the haptic device and virtual objects. With excessive delays, the penetration depth can be very large, which results in excessive forces applied by the haptic device. In this research, a new haptic network-tolerant force-feedback algorithm is proposed to deal with these excessive forces. A basic task of manipulating a virtual object along a track is reported, and the results suggest force feedback is adequate under the condition of the 100 ms delay; however, this system fails to handle the problem of distributed synchronization.

The delay is critical for applications that involve situations such as multiple users lifting a virtual object together. In [75], the results showed that when the round-trip is less than or about 300 ms, the synchronization control for the shared virtual object motion is effective; however, the haptic interaction is affected. At 120 ms, round-trip forces were somewhat oscillatory, and that led the participants to release the virtual object. This is one example of how a network delay may be acceptable in terms of visualization but not for realistic haptic feedback.

When collaborating in a distributed haptic assembly simulation, laboratory test experiments show adequate haptic interaction for delays of less than 60 ms [292]. For larger delays, haptic interaction in the case of collisions between the grasped objects may be affected. An experiment was carried out between Labein (Spain) and Queen's University, Belfast [182] with a 53 ms round-trip delay, and the results were satisfactory. In [203], during a cooperative session, the round-trip delay between MIT, Boston and UNL, London was about 90 ms. Various techniques for reducing the transmission delays have also been presented. For example, a damping factor and an algorithm for collision prediction were added.

Many authors believed that a latency of greater than 60 ms prevents usable collaborative haptics [340]. However, Shen et al. [341] found that, considering only solid objects, a 90 ms round trip latency was the threshold for haptic device stability. In addition, the authors in [47] have shown that with specialized physics, in surgical environments with soft objects, round trip latencies of 320 ms can be accommodated. This implies that locations on opposite sides of the globe can share haptic environments.

Network Jitter

In any network environment, there is no fixed delay for packet transmission. Every time a packet is sent, it can be received earlier than the previously sent packet.

The variation of the network delay is known as jitter. It has been shown that jitter has the greatest impact on coordination performance when the latency is high and the task is difficult [281]. From the user's perspective, the effect of jitter – with a fixed network delay – makes him/her feel that the object's mass is variable [158].

One way to smooth the jitter has been proposed by Gautier et al. [123] where messages are multicast to all participants who share a synchronized clock. A timestamp is attached to each updated message, and at the receiver side the processing of the received packet depends on their timestamps and not their time of arrival. This approach has two main advantages: the ease of implementation and the application independence. However, these advantages come at the cost of an increased overall delay.

Packet Loss

Packet loss occurs due to deficiencies in network operation and resources, which are mostly caused by network congestion (an overwhelming increase in the amount of traffic that goes into the network). Additionally, the queuing strategies that are implemented in intermediate routing nodes determine the conditions and the rates at which packets are discarded.

When dealing with position information, the loss of packets causes a small discontinuity in the object's position as felt by the remote collaborator. However, the loss of large bursts of position packets can easily impede and distract collaboration. Therefore, transport protocols prioritize transmitted packets and treat them differently based on their corresponding importance. For instance, packets that need reliability (usually named key update messages), such as a packet reporting a collision between the shared object and another object, are usually delivered using acknowledgment-based approaches (either positive or negative acknowledgment) to ensure reliable and timely transport. On the other hand, packets that do not require reliability, such as transient position information, are sent using pure best-effort services such as UDP.

The main effect of packet loss is that it minimizes the force feedback effect on the shared environment objects by reducing the intensity of the force [354]. Eventually, it leads to a desynchronized environment mainly by desynchronizing the objects' locations [158]. Furthermore, packet loss leads to considerable jitter, which in turn causes rebound or vibration forces that lead to instability.

Haptic Rendering

Not only is consistency a challenge, but providing reliable and compelling haptic interaction is as well. On the one hand, there needs to be fast computation for consistency so unstable haptic feedback is avoided [24]. On the other hand, the high update rates of haptic feedback along with network delays degrade the force rendering performance. The case of very precise synchronization exacerbates the problem of unstable haptic rendering. For instance, the case of two users manipulating the same deformable object, where each feels the influence of the other user on such an object, has been shown as a stable interaction [24].

When an operator is interacting with a remote environment, due to the network delay, a virtual object can be penetrated before its correct position is received, which may result in rebound or vibration forces [354]. In the case of great distances, such as in undersea and outer space operations, it is challenging to maintain stability in force feedback. Therefore, maintaining high update rates in the haptic servo loop, combined with the complexity of the sensory motor system, makes haptic rendering one of the most challenging issues to be addressed.

Latecomer Support

One of the major issues that applications in shared environments need to address is the ability for participants who arrive late to join an ongoing simulation. Existing approaches for latecomer support have been maturely investigated using audio and/or visual media and are usually by handled either the transport protocol or by the application itself (at the application layer). However, applying the same mechanism for haptic simulation has not yet been widely investigated and needs further research.

6.6 Architectures for C-HAVE

Fundamentally, there exist two network architectures for remote haptic collaboration, namely the client–server architecture and the peer-to-peer architecture, in addition to several other hybrid architectures (as elaborated in [243]). These architectures, particularly when incorporating haptic interaction, have two major concerns: consistency and responsiveness (as a result of network reliability, delays, and/or jitters).

Client–server systems provide high consistency; however, they suffer from low responsiveness. Due to the fact that haptic applications require high responsiveness, most collaborative haptic applications use the peer-to-peer distribution model. In order to compensate for lower consistency, several consistency control algorithms have been proposed to synchronize how the collaborative environment is shared and manipulated by distributed participants [248]. The goal in the design of HAVE applications is to find the best tradeoff between responsiveness and consistency. Furthermore, important factors shall be considered in the design process, such as object scalability, heterogeneity of haptic devices, and stability and safety of haptic interfaces. This justifies why many researchers are investigating the suitability of hybrid architectures for HAVE applications, where both architectures are combined in order to control the tradeoffs between consistency and responsiveness.

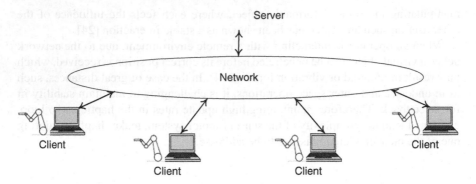

Fig. 6.10 The client–server architecture for tele-collaborative haptic applications

6.6.1 Client/Server Architecture

The client–server architecture is composed of a centralized server and one or more clients that are connected to the server via a computer network (as shown in Fig. 6.10). The server maintains the simulation of the shared haptic environment and updates the clients with environment changes. The clients receive the rendered simulation (graphic and/or haptic rendering) that they can view and interact with. Any interactions between the clients and the simulation environment (graphic and/or haptic) have to go directly to the server that updates the state of the simulation accordingly and eventually sends these updates to all the clients to update their local representations of the environment. As for haptic interactions, each client sends the time-stamped haptic device position and receives the interaction forces that should be applied by the haptic display (device). That is, the haptic rendering, including collision detection and force computations, takes place at the server side.

The client–server architecture's distinguishing feature is its ability to maintain consistency among participants, as the application state is stored and maintained by a central server [243]. Therefore, managing and updating the shared environment for all clients is straightforward, and usually synchronization of the client's view is not a major issue. Additionally, many researchers argue and have demonstrated that this architecture generates acceptable levels of latency and jitter [243]. Furthermore, the fact that the environment is completely controlled by a central server simplifies the start-up of applications, particularly if we consider latecomers joining the application environment. A latecomer does not need to contact every node in the network to inform them about their presence or to compose the state of the shared environment.

On the other hand, the client–server approach suffers from a few limitations. First of all, the architecture is characterized by limited responsiveness. That is, all interactions between any two clients have to go through the central server, which results in excessive delays and jitter are considered a major issue due to the fact that they cause haptic interface instabilities in haptic applications with strict communication requirements. This is not tolerated in most haptic applications. Second, scalability is another major issue in the client–server architecture since adding too many clients makes the server a bottleneck. Third, centralization means that the system has a single point of failure problem; if the server fails, the application will be inaccessible to all the clients. Furthermore, clients will not be able to communicate among each other because all communications have to go through the server. Finally, considerable client resources are being wasted because clients can provide plenty of processing and/or storage power of their own. The client–server architecture requires a high-duty server that is able to process complex computations and service all clients with minimum processing delays.

6.6.2 Peer-to-Peer Architecture

In the peer-to-peer architecture, every participant (or peer) maintains a local replica of the HAVE application with which the client interacts and views. Therefore, the interaction with the HAVE application is happening locally, so it is considered of high responsiveness. However, each peer should communicate its interactions and updates with all other peers in the network to maintain the consistency of the environment view across all network peers, as shown in Fig. 6.11. The major advantage of the peer-to-peer architecture is high responsiveness since there is no interaction between the client and a server. This is the reason why several HAVE applications adopt the peer-to-peer architecture [181, 308]. Furthermore, peer-to-peer architecture makes better use of the distributed client's resources due to the distribution of processing and storage load across the network peers. In a typical peer-to-peer network configuration, the following interaction procedure occurs:

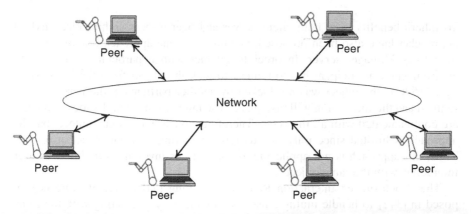

Fig. 6.11 The peer-to-peer architecture

- Each peer runs and maintains the simulation locally and interacts through the haptic device with the end user.
- As a result of user interaction, the client updates the virtual environment, calculates the reaction forces locally, and renders the interaction forces directly to its end user. The client might incorporate any received update messages from other peers into the haptic and/or graphic rendering of the local view of the HAVE application.
- The client broadcasts an update message to all the network peers to update their local views according to the client's interaction.
- Other peers receive and handle the update messages by updating the state of the application according to the new updates from all other peers and generate a new local state of the application. Notice that for realistic haptic interaction, the update rate should be as high as 1 kHz, which is achievable in this case because the haptic rendering loop is completely local for every peer.

Responsiveness and scalability are the distinguishing features of peer-to-peer architecture, and that is why it is the most commonly used architecture in haptic applications. However, maintaining the environment consistency is not a trivial task since the state of the environment is computed by each client individually and eventually delivered to the peer clients. Furthermore, the peer-to-peer scheme consumes significant network bandwidth due to the huge traffic generated by clients communicating their states with other participants. Finally, the procedure for latecomers joining the application is not as intuitive as in the case of the client–server architecture. Latecomers need to communicate with all other peers in the network to compute the current state of the environment and generate a consistent view of the application.

6.6.3 Hybrid Architecture

To inherit benefits from both client–server and peer-to-peer architectures, hybrid approaches have also been proposed. For instance, one approach requires the use of a Lock Manager server. In order to interact with a particular object in the environment, a participant has to request to lock the object from the Lock Manager [159]. If the object was not locked by another participant, the permission is granted and the interaction will take place. The interactions with the locked objects are communicated with all the peers. Therefore, consistency is guaranteed, but the approach is limited since only one user is interacting with an object at a time. Thus, this approach is not applicable to applications involving simultaneous user interaction with the same object.

The Synchronized Interaction Request Resolving (SIRR) architecture is proposed in [342] to handle multiple users simultaneously interacting with the same object. In the SIRR, when multiple users intend to co-touch a common object in the environment, a central server (called the interaction manager) is used temporarily to organize such an interaction. The central server maintains and manages the state of the shared object during the interaction. The new state of the object is then communicated to all the participants.

Another hybrid architecture, proposed by Marsh et al. [243], is to use a roaming-server. This means that a server that acts as a simulation engine is strategically placed at each local area network. Clients connecting to the system will choose their closest simulation engine (the one residing on their local network). Therefore, the client–server approach is adopted within the local area network to keep delays minimal, and the peer-to-peer approach is adopted for the inter-server communication (across different local area networks). Consistency is easily ensured within the same local area network. If a short period elapses with none of the users interacting with any object in the environment, all the servers are replaced in an inactive state [159].

6.7 Communication Frameworks for C-HAVE Systems

The problem of communicating multimedia data that incorporates haptic data has been the subject of research for the last decade. There have been three branches in handling this research subject: (1) improve the control mechanisms at either ends of the communication to accommodate the unpredictable behavior of the network, such as through the use of delay compensation techniques [158], jitter smoothing algorithms [281], or haptic data compression (application level solutions), (2) designing novel transport protocols or adapting existing ones for haptic data communication (transport layer solutions), and (3) designing statistical multiplexing communication frameworks (application level solutions) to handle the communication of multiple media, including the haptic media.

6.7.1 Compression and Control

As previously discussed, the data for nonlinear haptic media that stores haptic surface properties, etc., is small in size, and once it is transmitted, it does not need to be transmitted again until the properties are changed. Linear haptic media, such as position, velocity, acceleration, force, torque, etc., is much smaller, but its transmission rate must satisfy the 1 kHz critical condition. Since its update rate is very high compared to audiovisual media, its bit rate (the number of bits that are conveyed per unit of time) is also quite high, so linear haptic media needs to be compressed.

In the early stages of haptic media compression, a stochastic approach was exploited based on the fact that the changes between each sample value of linear haptic media, namely position, velocity, and force, are considerably small. With a Differential Pulse Code Modulation (DPCM)-based compression scheme data can be significantly compressed by reducing the required number of bits [157]. This scheme transmits the initial value, and then only the differences are quantized and transmitted. At the receiver side, a current value can be restored by adding the current difference value and the previously stored value. Since the difference values are much smaller than the original value, the number of bits needed to quantize the transmitted values can be reduced. However, packet loss can substantially affect the very sensitive linear haptic media, therefore, the exact value should be sent within a certain time period. Shahabi et al. [337] extended this work and adopted ADPCM (Adaptive DPCM); the difference values get bigger as one moves the haptic device faster, so the quantization step size becomes larger to accommodate this fast motion.

Statistical approaches help to reduce the whole amount of data to be stored or transmitted, especially when there are many haptic devices involved in an application and a huge amount of data needs to be dealt with. However, if real-time applications such as tele-operation and tele-presence systems are taken into account, the 1 kHz update rate requirement for better fidelity and stability makes it less efficient in terms of the amount of data, and even maintaining the stability of the application. For example, 3-DOF velocity data can be represented by 12 bytes by assigning 4 bytes to each component. If this data is transmitted through the Internet, UDP/IP can be used to carry the data, and the header size is 24 bytes per packet (20 bytes IP, 4 bytes UDP). Because of the 1,000 Hz update rate, every sample of the velocity data needs to be immediately transmitted, and the network resources end up being abused by the overhead headers of the network protocol. This motivated the advent of the perception-based deadband approach [164].

Based on Weber's law, the perceptible difference of stimulus intensity is a well-known psychophysical finding that is represented as a proportional relationship with the size of the base stimulus intensity. For example, consider a person holding a weight of X grams as a reference. By gradually adding weight, they would then only perceive a weight difference once a weight of Y grams is added. If the reference weight is doubled to $2X$, $2Y$ grams is needed to notice the weight difference. It can be written as:

$$\frac{\Delta I}{I} = k,$$

where I is the base intensity of stimulation and ΔI is the added intensity required for the difference to be perceived, called the just noticeable difference (JND). The constant is called the Weber constant and varies with stimulus types and receptors.

In this approach, a sample of haptic media is sent first, and another sample is sent whenever the intensity difference exceeds the JND. If the difference is not perceptible or is lower than the JND, the current sample is dropped and not transmitted. On the receiver side, during the time interval where no new packet arrives, a modified 'hold last sample' estimates the current sample based on the time interval and the last sample received while considering the passivity/stability of the whole system. This method was applied to a tele-operation system that used a 1-DOF haptic device in [164]. The measured velocity was transmitted from OP (operator) to TOP (tele-operator), and the measured force was transmitted from TOP to OP for the force-feedback control loop. They could set the Weber constant to 10% with subjects barely noticing or not being disturbed at all, and the packet rates of velocity and force were reduced by 25% (velocity) and 5% (force) of the original rates of 1,000 Hz. The modified "hold last sample" to guarantee the stability is further introduced in [166] and [167]. The stability issue is further investigated in [216].

This approach was then applied to 3-DOF cases by introducing a sphere-shaped dead zone [161, 163]. In this case, the sample difference was represented as a 3D vector, and the sample was transmitted only when the difference vector was outside of the dead zone. They could achieve 75 and 90% reductions of the packet rates with Weber constant values of 20 and 5%, respectively, with subjects barely perceiving the degradation of the interaction quality.

The packet rate reduction could also be increased by adding a prediction algorithm. The predictor predicts the next sample of haptic media by extracting the predicted value from the current sample. Since the same predictors are used on both sides of the tele-operator system, the receiver side uses the predicted value when packets do not arrive [162, 199, 407]. However, the prediction can be substantially degraded when the signal is affected by noise, especially for velocity predictions that are derived from the position data that already contain noise. Hinterseer et al. [165] adopted a Kalman filter to reduce the noise level before prediction is applied. They could considerably reduce the packet rate of velocity while ensuring that the force was not affected significantly.

Furthermore, Zadeh et al. [406] investigated how the velocity of a hand movement affects the force JND. In their preliminary experiment, they found that for low velocities ($0.03–0.05 \, \text{m s}^{-1}$) the average force JND opposed to the movement is 51 mN, and for high velocity ($0.22–0.28 \, \text{m s}^{-1}$) it was 97.7 mN. The aid force (force in the same direction as the subject's hand movement) JND was 49.6 and 89 mN, respectively. Based on this finding, while the user's hand is in motion, the force JND increases so that there is more room for increasing the compression rate. Kammerl et al. [200] modified the previous perception-based compression scheme based on this finding by using a modified Weber's law:

$$\frac{\Delta I}{I} = k + \alpha J,$$

where J is a dependent stimulus intensity that affects the base stimulus intensity I, and α is a constant. They applied this to velocity and the force haptic factors and obtained the size of the applied velocity-adaptive deadband bounds with:

$$\Delta_i = (k + \alpha \cdot |\dot{x}_i|) \cdot |f_{i-m}|,$$

where $|\dot{x}_i|$ and $|\dot{x}_i|$ are the velocity and the force, respectively. Also, i is the current time and m is the time when the last packet was transmitted. In their experiment, they could achieve an additional data reduction of up to 30% compared to the previous Weber-inspired approach without perceptibly impairing the quality of the force-feedback interaction.

6.7.2   Transport and Network Protocols

Only a few communication protocols have been designed for multimedia applications since it is difficult to capture the widely varying requirements of different media into one generic protocol. The generic transport-layer protocols (namely Transport Control Protocol (TCP) and User Datagram Protocol (UDP)) are used by several multimedia applications. Only a few protocols have been proposed and evaluated for haptic applications (such as SCTP, Smoothed SCTP, Light TCP, RTP/I, and STRON).

TCP provides several services that have a negative impact on haptic applications, including error control, sequence control, loss control, and duplication control. If a haptic update is lost during transmission, all the updates that follow will be needlessly buffered on the receiver side while network resources are being exhausted in order to re-transmit an obsolete one [98]. UDP does not suffer from any of the drawbacks of TCP, however, it does not fully suit the reliability requirements of haptic applications [98]. For instance, UDP has lower jitter and no buffering delays because it does not resend lost packets [98]. Nonetheless, UDP remains the most popular transport layer protocol for real-time applications and has consequently been used as the transport protocol for many haptic data communication instances.

The Synchronous Collaboration Transport Protocol (SCTP) is similar to the UDP protocol in that most of the messages are sent unreliably. What differs is that key messages are sent reliably. It also differs from UDP in that sequence numbers are used for packet ordering. For each key message, a timer is set, and if a timeout occurs before receiving the acknowledgement of the message in question, it is resent as a key update [347]. It has been proven that for collaborative applications, SCTP performs better than protocols with negative acknowledgments [347].

The Smoothed SCTP protocol adds a jitter smoothing mechanism to the SCTP protocol. On the sender side, the Smoothed SCTP and regular SCTP protocols are the same. However, on the receiver side, each received update is placed in a "bucket" according to its timestamp. The receiver constantly checks for updates that have been sent at a constant rate (measured in milliseconds), and in the case that an update is found, it is retrieved from the "bucket" and forwarded to the application [98]. This means that all updates, including the ones generated locally, are processed with constant delays (δt).

Light TCP is inspired from TCP and supports the concept of message obsolescence. The sender queue accepts update messages from the application and processes them as follows [98]: (1) a key update is placed at the end of the queue and marked as a key message in order to prevent it from being erased, (2) a normal update can replace older normal updates for the same shared object, and (3) unacknowledged update messages are placed back in the queue if no newer updates from the same object have been produced by the application. At the receiver, a received update is immediately forwarded to the application if its sequence number is bigger than the last received update's sequence number, otherwise it is dropped. There is no buffering.

The Real-Time Protocol for Interactive applications (RTP/I) is an application layer protocol designed for general network distributed interactive applications, and haptic applications fall under this category. The communicated messages are categorized as: event, state, and request-for-event packets [248]. An event packet carries an event or a fraction of an event. A state packet carries the entire state of a subcomponent in the environment. The request-for event packets are used by participants to indicate that the transmission of a subcomponent's state is required by the sender [248].

The design of transport protocols that optimize data transmission for HAVE interactive applications is not well explored. A few approaches exist, such as the Real-Time Network Protocol (RTNP) [378], Interactive Real-Time Protocol (IRTP) [289], and Efficient Transport Protocol (ETP) [395], but with limited results. The first model includes a priority mark in the packets but still leaves the network jam problem unresolved, and IRTP is not widely developed. ETP aims at optimizing the available bandwidth within a network so that the highest number of packets is sent without affecting each packet Round Trip Time (RTT). In ETP, the device controller is aware of the actual RTT at any time. It is an important issue in C-HAVE systems since, if properly designed, the haptic device controller can make decisions to counteract the effect of communication delay.

The research presented in [65] proposes a framework of Quality of Service (QoS) management for supermedia tele-operation systems. In this work, latency-sensitive supermedia streams are encoded using redundancy codecs and transmitted over multiple overlay paths. The overlay routes and encoding redundancy can be dynamically tuned to meet the QoS requirements of the supermedia streams to compensate for network performance degradation.

The authors in [216] proposed a haptic data transport scheme that reduces the transmission rate by using adaptive aggregated packetization and priority-based filtering. The proposed scheme adapts the transmission, loss rate, and buffering time of haptic events to the current network state based on the delay and loss effects of each haptic event. In order to offset jitter with a small playout delay (delay to handle received packets and eliminate jitter), an intramedia synchronization scheme with dead reckoning is used. The priority-based haptic event filtering and network-adaptive haptic event aggregation of the proposed transport scheme have resulted in a lower transmission rate than the other transport schemes (Reed–Solomon FEC error control, selective ARQ error control, and congestion control).

ALPHAN [12] is a protocol that uses a similar approach, but it uses a multiple buffer scheme to prioritize and optimize media data transfer. Additionally, ALPHAN uses the HAML description language [102] to define the application requirements and pass them on to the network protocol. The protocol supports the notion of key updates, which is widely supported by most of the haptic communication transport layer protocols. This is done by implementing an application layer reliability mechanism that is only applied to key updates, while normal updates remain unaffected. ALPHAN also makes use of the Multiple Buffering (MB) scheme. In this scheme, every object in the application is attributed a sending buffer. Allocating a buffer for each object permits the decoupling of update transmissions for different objects, especially if they are independent from each other or they need to be prioritized based on user and/or application preferences.

Figure 6.12 shows a high-level view of the ALPHAN architecture. The Protocol Manager Module connects all the components and interfaces directly to the application. The Simulation Module holds both the graphic and haptic simulations. The Sender Module comprises the sending buffers. The updates that are placed in the buffers are sent as soon as possible according to their priority. The Receiver Module is responsible for receiving updates and handling them. If a key update is received by this module, an acknowledgment is produced and buffered in the acknowledgment queue where it is retrieved by the Sender Module. All the updates received by the Receiver Module are buffered in the Update Relayer buffer. The Update Relayer Module is responsible for relaying packets to the Protocol Manager at the appropriate time (which is inscribed in the timestamp of the update). Such a mechanism is necessary for the implementation of the local lag algorithm for consistency purposes or other jitter smoothing algorithms. The Network Sensing Module is responsible for calculating the Round Trip Time (RTT) value and detecting any disconnections. The Network Time Protocol (NTP) Module is used to synchronize the simulation timer with the timers of the other participants' simulations.

6.7.3   Statistical Multiplexing Schemes

Statistical multiplexing is a proven technique used to improve the efficiency of communication over a limited bandwidth network [73]. The principle is that a group of media channels share a limited bandwidth that is allocated on a frame-by-frame basis by a centralized controller (the multiplexer). This ensures that the channels with the most demanding QoS requirements are allowed to borrow more bandwidth than channels with less QoS requirements. Most statistical multiplexing research is done at the transport layer (especially ATM networks [73]). At the application layer, there has been very little effort in using adaptive multiplexers for communicating multi-modal data that includes haptic media [337].

One of the few works in this area involves dynamically controlling the arrival rate of multimedia data by switching the coders to different compression ratios (changing the coding rate) based on the network conditions [327]. The technique was tested with audio media, and it was found that the link performance is

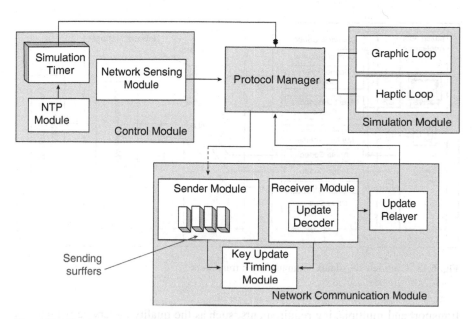

Fig. 6.12 High level view of ALPHAN architecture

significantly improved in terms of reducing the probability of call blocking and enhancing the multiplexer gain.

The work in [408] investigated the use of self-organizing neural networks to design a statistical multiplexer for video streams. The proposed approach uses multiple video coders followed by a multiplexer that generates the aggregate sequence for several video streams. Three control methods are proposed: priority control, rate control, and a combination of both. The neural network approach performed very well in improving the packet loss, as compared to the Round Robin (RR) approach [304], and in smoothing the variations of delays using rate control.

Lately, another application layer communication framework for a synchronous haptic-audio-visual communication framework, named Admux, has been proposed in [104]. The authors propose to use a statistical multiplexing scheme that is adaptable to both the application requirements and the network changes. Admux uses multiple channels that enable it to enforce media prioritization since each channel can be treated differently by the multiplexer. For example, the haptic channel can be assigned a higher priority level than the audio and video channels. Finally, Admux is based on UDP, which means it is Internet-based. An overview of the Admux communication framework is presented in Fig. 6.13.

The application generates multiple streams of media data. First of all, these streams are compressed using different codecs (depending on the media type), and then the compressed streams are multiplexed using the Mux block (Fig. 6.13). Based on the available network resources, the multiplexer dynamically re-configures the codecs to comply with the available resources. The HAML-QoS defines the

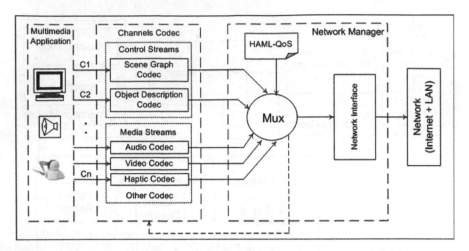

Fig. 6.13 Overview of Admux communication framework

transport and multiplexing requirements, such as the quality of service parameters for each input channel, the multiplexer configuration, etc. It also contains the number of input channels, their respective network requirements, and the associated codec configurations. Finally, the network interface packetizes and transmits the multiplexed stream using a particular underlying transport protocol (UDP in the case of Admux). The inverse of this process is performed at the receiver side when the received data is de-multiplexed and forwarded to the corresponding destination channels and eventually displayed using the appropriate interface.

6.8 Quality of Experience in Multimedia Haptics

Quality of Experience (QoE) is the ultimate decisive factor in the success and popularity of a certain technology. If the technology has a high QoE, then it will be worth investing in, otherwise the cost will seem to outweigh the benefits of the technology, and it will ultimately fade away [392]. User centric studies focus on the concept that what really matters in any measurement attempt of quality is the user. Quality of Service (QoS) can help determine the satisfaction of the user, but it is not guaranteed that a high QoS will lead to a high QoE.

Measuring the QoE of an application or of a technology is not a straightforward task. User reasoning includes many parameters and factors, so it is not easy to quantify their experiences. The haptic domain is no exception, however, there is research in progress to determine the QoE of haptic applications.

It is essential to establish some guidelines for determining and improving the QoE of haptic-based applications. Two concerns are:

- Will haptic hardware dramatically change in the future? (this might affect the QoE)
- Will there be any side effects from the fact that there are no reported cases of haptic devices used in everyday life for a prolonged period of time?

QoE is gradually becoming an important measure for the evaluation of multimedia applications. As Jain [191] puts it, we require improved performance measures over the well-established QoS measures to deal with the subjectivity of the user. The relation between QoS and QoE has been addressed in [397]. Instead of extending the QoS metrics, the paper relates the performance measures of the QoS to QoE measures according to quantified correlations. The result is a theoretical framework for computing QoE using both QoS and QoE metrics.

There has also been some work done in evaluating virtual environments. The evaluation methods and the aspects to be evaluated vary depending on the type of the application and the parameters to be evaluated. Basdogan et al. [29] conducted studies to evaluate the haptic feedback role in collaborative human–human and human–machine interactions in shared virtual environments (SVEs). The evaluation consisted of measuring response variables as well as giving questionnaires to the users undergoing the experiment. Another approach to measure haptic benefits is given in [141]. The authors directly measure physical parameters generated by the haptic device in order to assess the quality of the application. They suggest that this is a complementary approach to conducting a statistical survey after users test the application. Some of the parameters that they chose to include in their physical survey are gesture position and gesture velocity.

In [392], the authors discuss some of the methods and challenges in determining performance measures in the context of virtual reality applications. They indicate that there are three ways of assessing QoE performance measures: subjective ways through interviews and questionnaires; task performance measures through observation of the user; and a physiological approach via biological indicators such as heart rate. Taking stress as an example, there are direct measurements that can indicate if the user is stressed under prolonged exposure to a virtual environment. Under stress, the sympathetic nervous system is activated, and blood volume, heart rate, and respiration rate all increase. Ramsey [299] argues that measuring those symptoms directly is more effective than a questionnaire due to three limitations:

1. People are mentally aware of their internal state (emotional condition) when, under the same circumstances in the real world, they would normally not be. For example, users might experience stress or fatigue without being mentally aware of it.
2. People might not understand the implication of the response in the questionnaire.
3. People may not wish to report feeling any symptoms.

In the following section, we have attempted to collect possible parameters for QoE evaluation of multimedia applications, including haptics applications.

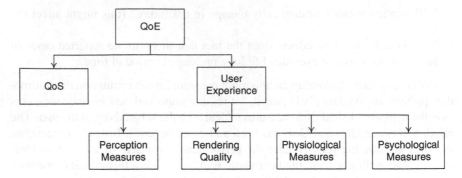

Fig. 6.14 Higher level organization of QoE model

6.8.1 Quality of Experience Model

In this section, we present an example QoE model and the taxonomy used to organize the different parameters contributing to QoE for C-HAVE applications [145]. The taxonomy is based on subjective vs. objective metrics. As such, the parameters are divided into two groups: ones that can be measured directly from the application, such as forces and delay, and those that must be deduced by other means, such as a user questionnaire or behavior (like intuition). A different taxonomy stems from the core definition of QoE as defined by Jain [191], where the top organization level is comprised of two parts: the QoS and the user experience. The user experience can be further subdivided into four parts: perception measures, rendering quality, physiological measures, and psychological measures. This higher level organization, shown in Fig. 6.14, reflects an apparent taxonomy for a virtual reality applications evaluator, and at the same time, is more customizable depending on the parameters needed for the evaluation [145]. As an example, developers wishing to evaluate only the QoS of the application can disregard the user experience parameters.

6.8.1.1 Quality of Service Parameters

QoS parameters ensure the smooth flow of the application for the user or, in certain cases, the customer. Most parameters are standard for any networked application but looking at synchronization, it can be divided into two parts: network synchronization, which is common to network applications, and media synchronization, which is specific to the multimodal side of haptic audiovisual environments. The following is a definition for the most common QoS parameters:

- *Response time*. The time taken by a system to respond to an action. It is measured in milliseconds or microseconds.

- *Latency/Delay.* Time taken for the packet to reach from source to destination. It is measured in milliseconds or microseconds. From the source to the destination. It is measured in milliseconds or microseconds. The different sources of delay are: (1) propagation delay, which is the delay through a physical medium, (2) link speed, which is determined by the link bit rate, (3) queuing delay, which represents the time spent in router queues, and (4) hop count, where each traversed router or switch adds queuing delay.
- *Price.* The quantity of payment or compensation given from one party to another in return for goods or services. It can be measured by a metric related to energy, money, automation, or other efficiency of the service.
- *Privacy.* Deals with what personal information can be shared with whom and whether messages can be exchanged without anyone else seeing them.
- *Security.* Is defined as the level of protection for the information exchanged through the use of multimedia technologies.
- *Availability.* Is defined as the ratio (or probability) of time a system or component is functional to the total time it is required or expected to function. Small probability values for availability indicate bad QoS, while high values indicate good QoS.
- *Bandwidth/Throughput.* It is the amount of data transferred from source to destination or processed in a given amount of time. Measured typically in bits/second or bytes/second.
- *Network Synchronization.* Refers to the temporal relations linking the various media objects within a multimedia presentation. Example: Time relations of a multimedia synchronization that starts with an audio/video sequence, followed by several pictures and an animation that is commented by an audio sequence and haptic feeling.
- *Media Synchronization (intra-modal).* Refers to the temporal relations between media units within a time-dependent media object. For a video with a rate of 25 frames per second each of the frames has to be displayed for 40 ms. For haptic data with 1 kHz, each of the data samples must be captured and displayed for 1 ms.
- *Jitter.* Difference in latency of network packets, usually measured in microseconds or nanoseconds.
- *Reliability.* Is defined as the ability of the computer system and its components, i.e., a haptic audiovisual environment to consistently perform according to the given specifications.
- *Error.* Sometimes C-HAVE packets are corrupted due to bit errors caused by noise and interference. The receiver has to detect this and, in case the data contained in the packet is needed, may ask for this information to be retransmitted.
- *Safety.* Defines the aspects to be considered in order to operate the haptics environment properly and use it in conjunction with other peripheral equipment without damaging the environment and the users.

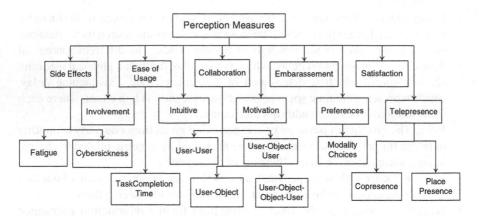

Fig. 6.15 Perception measures parameters

6.8.1.2   User Experience

The second part of QoE is the user experience. This is an important evaluation category for the overall quality of the application. Even if the application possesses excellent QoS parameters, users might still feel that the application is not up to their personal standards for some reason. The application might not be exciting enough, not easy enough to use, or may cause dizziness, which is referred to as cybersickness.

Perception Measures

As depicted in Fig. 6.15, perception measures mirror how the user perceives the application. This is a user-centric category, and could be unique for every user. Some users may get tired from the application, while others may feel relaxed. Some might feel the effect of collaboration in a C-HAVE, while others might need more stimuli. Each user may have a certain set of preferences and modality choice.

Another point to consider is the fact that there are different levels of experience among users. While a certain group of users could be very experienced with virtual reality applications and very dexterous using haptic devices, others may be novice users and less skillful. This variation in the level of experience will cause users to have different perceptions regarding the application. When evaluating a HAVE application, it is essential to include different categories of users and to ensure that the application suits a wide range of audiences.

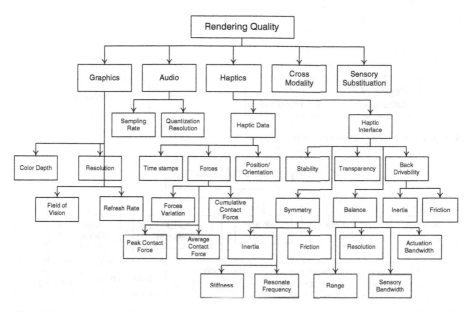

Fig. 6.16 Rendering quality parameters

Rendering Quality

The rendering quality relates to the quality of the three major modalities, namely graphics, audio, and haptics. Each modality is evaluated separately at first, and then eventually blended and mixed modalities are evaluated. As seen in Fig. 6.16, there is an emphasis on the haptics modality since it has very stringent requirements in terms of feedback loops, which might affect the stability and transparency of the application.

Physiological Measures

Physiological measures are biological parameters that are measured directly from the user's body while they are using the application. These parameters directly determine factors such as cybersickness, stress, and brain activity (Fig. 6.17).

Psychological Measures

Unlike the physiological measures, psychological measures reflect the status of the user through observation and not direct measurements. Observation can assess the psychological behavior of users, such as stress, without hindering the user

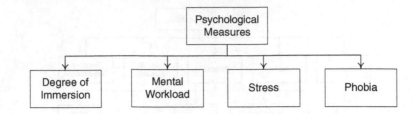

Fig. 6.17 Physiological measures parameters

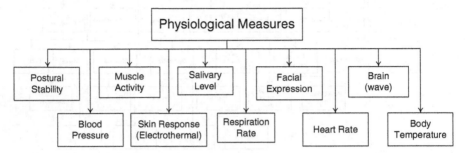

Fig. 6.18 Psychological measures parameters

movements by including measuring devices. Psychological measures are displayed in Fig. 6.18.

6.9   Haptics Watermarking

With the wide availability and widespread use of different media in several fields, such as entertainment, the medical industry, the military, etc., and with the recent advancements of haptic technology, it is expected that in the very near future the need to protect haptic-enabled virtual scenes and environments from malicious attacks or inadvertent tampering will arise. In recent years, digital watermarking, which deals with the process of embedding information into digital data in an inconspicuous manner, has been proposed as a viable solution to the need of copyright protection and authentication of multimedia information. Example applications of digital watermarking include identifying the origin, owner, use, rights, integrity, or destinations of multimedia content (e.g., digital images, video, audio, and 3D models).

Watermarking techniques are different from other intellectual property rights methods in that they are imperceptible, they undergo the same transformations as the content, and they are inseparable from the digital content in which they are embedded. In fact, the first requirement of digital watermarking techniques, regardless of the addressed media or application, is imperceptibility. This refers to the perceptual similarity between the original and watermarked data. Ideally, the perceptual quality of the watermarked media must be identical to the original.

Recently, considerable progress has been made in 3D watermarking where the main focus has been on triangle meshes, the most common digital representation of 3D models due to its simplicity and usability. Existing watermarking techniques concerning 3D meshes host the watermark either by modifying the geometry or the connectivity of the surface (spatial domain) or by modifying some kind of spectral-like coefficients (spectral domain). Accordingly, the watermark's intrusiveness can be evaluated in terms of its visibility in the rendered version of the mesh. The evaluation of 3D watermarking algorithms against the imperceptibility constraint has been thus far exclusively based on the sensitivity of human vision to distortion. Moreover, currently available perceptual metrics generally used to assess the quality of watermarked 3D meshes have been validated solely through psycho-visual experiments [86, 312].

Very recently, however, studies have been conducted by Sakr et al. [323–325] to investigate the role of multisensory feedback in the perception of a watermark embedded in a haptic-enabled 3D virtual surface. The authors investigated the following research questions: Is the haptic sensory channel more sensitive than the visual sensory channel in detecting a watermark embedded in a haptic-enabled 3D object? Do watermarks inspected using multimodal feedback (vision + haptic) result in very different detection thresholds from those detected using a single sensory modality (touch-only, or vision-only)? Or more importantly, does visual feedback, when presented together with haptic feedback, improve the perception of a watermark embedded in a 3D mesh?

Sakr et al. [323–325] argue that while it is intuitive to assume that a multimodal presentation of stimuli should lead to an improvement in performance (e.g., for watermark detection), previous research in human perception suggests otherwise. Specifically, the role and possible advantages of multisensory feedback in roughness perception (the 3D watermarking process can be regarded as a particular form of surface roughness) is quite complex as it can vary across different experimental conditions, including the type of surface, haptic device used, surface parameters, etc. The experiments were performed using a visual–haptic interface that enabled users to see and touch virtual objects at the same location in space. A detailed analysis of the results was conducted to statistically explore the impact of the considered modalities on the measured watermark detection thresholds across different resolutions of the underlying virtual 3D mesh.

Overall, the results suggested that "haptic-alone" is superior to "vision-alone" in detecting a watermark embedded in a 3D mesh; however, relying on bimodal visual-haptic feedback is better than any of the single modalities. In addition, it was assessed that the impact of the selected modality on the perceptibility of the 3D watermark is independent of the chosen surface resolution. The work by Sakr et al. [323–325] was a very important first step toward the analysis of multimodal visual–haptic watermarking. The authors' findings are expected to stimulate the reevaluation of existing mesh watermarking algorithms (using a vision–haptic setup) and will serve as a basis for further studies in haptic digital watermarking.

6.10 Closing Remarks

Haptic applications are increasingly and subtly being used in our daily life activities in the form of vibrating phones, joysticks, game controllers, and force-feedback controls. The next emerging idea will be pertaining to immersive tele-haptic environments where multiple users can interact with each other as well as with other digital media (i.e., 3D graphical models, video, images, etc.) by means of touch, as if a real world was in front of them. This research can have a powerful impact on the development of a new breed of human–human interactions.

Even though several candidate standard proposals have been introduced by the research community (such as HAML, X3D, GOTHI-05), there is no internationally accepted standard for haptic interactions and representation. One of the foreseeable efforts is the adoption, by ISO for instance, of a standard proposal that provides guidelines for haptic interactions and standard representations of haptic data and systems.

The communication of haptic media remains a major challenge in the research community. The strict QoS requirements for haptic applications (such as a delay of 1 ms) are very hard to meet, especially for nondedicated networks such as the Internet. For many C-HAVE applications to gain higher acceptability and popularity in the general public, these applications should be used with the Internet network because it is the cheapest and most acceptable network for public use. Therefore, finding a prominent solution for haptic data communications remains a major impedance to the proliferation of C-HAVE applications and systems.

Finally, research is being done to investigate the contribution of haptic modality to the overall quality of experience for end users. The fundamental question is whether the incorporation of haptic modality in multimedia systems would enhance the overall quality of user experience. Furthermore, questions can be raised as to how the advantages of enhanced senses outweigh the costs of adapting to a new technology. Moreover, how will users experience such advantages, and will they be overwhelmed or exhausted from interacting with multiple media?

Chapter 7
Touching the Future: HAVE Challenges and Trends

7.1 Introduction

Many researchers insist that the field of haptics is still in its infancy. Most of the challenges discussed in the next sections are hot research topics in this field, and some of them are currently being studied. Advances on each of these described fronts are certainly to be expected in the near future. As the prices of haptic devices drop, gaming industries will be the first ones to take notice and exploit the technology to complement their playing consoles, thereby creating a significant advantage over their competitors. Home rehabilitation applications will most certainly follow. Governments are desperate to reduce healthcare costs, so development of practical applications of this sort would be encouraged and funded. In addition, haptic technology will undoubtedly gain an important place in the education sector. New educational approaches rely on visual and auditory demonstrations to describe unintuitive concepts to students; the incorporation of the sense of touch is simply a natural progression of this trend to newer media. The field of networked haptics will be the last to see any considerable advances. This is mainly due to the fact that it relies heavily on the state of the network. Haptic applications have stringent QoS requirements that most nondedicated networks cannot guarantee. Until these network infrastructures are upgraded, networked haptic applications will continue to struggle from the lack of QoS guarantees. Last but not least, increased computing power will allow for the implementation of complex, multi-point, physics-based haptic rendering algorithms to support more natural interactions with a virtual or remote world.

7.2 The Golden Age of Haptics

There has been intensive research and development in HAVE technologies in the last few years, and some of these technologies have been commercialized successfully in certain areas, such as medical, gaming, and military fields [109].

A. El Saddik et al., *Haptics Technologies*, Springer Series on Touch and Haptic Systems, 183
DOI 10.1007/978-3-642-22658-8_7, © Springer-Verlag Berlin Heidelberg 2011

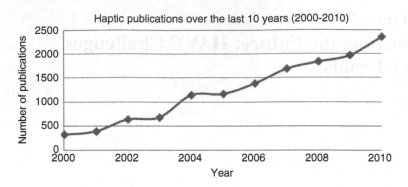

Fig. 7.1 Haptic publication trend over the last 10 years (2000–2010)

The da Vinci minimally invasive surgery system is a good example of success-ful incorporation of haptics technologies in the medical field. In the meantime, companies have been manufacturing 3-DOF and 6-DOF force feedback devices, and the price of the devices has become reasonable. In addition, high-quality open-source haptic rendering APIs are available on the Internet, and a variety of haptic stimuli can be displayed through them. In recent years, with the advent of hand-held devices, finger touch has become the main form of interfacing, and tactile feedback on the finger and hand has become the intuitive option for consumers.

Many of the fundamental challenges, such as haptic rendering and actuation technologies, have made magnificent progress in the last 10 years, however, some major issues remain unresolved. We conducted a survey of haptic-related publica-tions from the last 10 years (January 2000 – December 2010) in major publishers' databases to examine research trends. These databases were the Institute of Elec-trical and Electronics Engineers digital library (IEEE Xplore), the Association for Computing Machinery (ACM) digital library, and the Springer Publishing database. We found that the number of published research papers has steadily increased over time, as shown in Fig. 7.1. This demonstrates that haptic research has witnessed increasing interest by the research community. In addition, the last 2 years have witnessed a significant increase in the number of publications specifically related to tactile feedback as applied to pervasive devices, marking a potential trend in haptics applications.

Haptic technology is on a roll; more than 20 smart phone models have already adopted tactile capabilities to enhance the user experience (including the Nokia N8 and Samsung Galaxy S series). However, much more is yet to come. In the next sections, we provide thoughts about where haptic technologies are heading and potential research avenues in this domain.

7.3 Human Haptics

One of the fundamental research areas that justifies the integration of a haptic modality in multimedia systems is multi-modal cognition and psychophysics. Even now, our understanding of how the body and mind perceive and respond to multi-modal stimuli is incomplete. While unimodal cognition and psychophysics are much better understood, there are relatively few studies of multi-modal integration due to the difficulties in performing such experiments. Understanding how the brain perceives and responds to multi-modal media is a field that is still in its infancy. The following subsections describe some research and development aspects related to human haptics that we consider important in the HAVE field.

7.3.1 Human Perception and Quality of Experience

Several usability tests have proven that the insufficiency of information in haptic technology impedes the possibility of full tele-presence perception. Users suffer from a lack of in-depth information (the depth dimension) and poor tracking of the relative position and orientation of the remote participants. Another perceptual issue is the sense of space in the HAVE interactions. The issue here is that users apply previous real-world interaction experience in a simulated virtual environment that deviates from such realities and eventually creates confusion. The same applies for the perception of time. Due to the unavoidable delays in networked HAVE applications, there will always be a drift between remote collaborators. This deteriorates the participant's perception of time and creates further synchronization issues, particularly for collaborative tasks. Consequently, measuring the Quality of Experience (QoE) of HAVE applications is not a straightforward task, especially since user reasoning includes many parameters and factors that are not easy to quantify. The haptic domain is no exception, however, as described in Chap. 6, there is research in progress to determine the QoE of HAVE applications. It is still essential to continue further investigation in this field and to establish some guidelines for determining and improving the QoE of HAVE applications. At the same time, there are no reported cases of haptic device usage in everyday life for prolonged periods of use. Therefore, studying the side effects HAVE application immersion is a challenging topic that warrants investigation and research (such as the social implications). Another human-related haptic research trend is how humans communicate haptic stimuli. The association between what a user feels versus what they perceive still requires further studies and analysis. In particular, learning people's ability to distinguish and associate complex haptic stimuli is a must for designing efficient and highly usable haptic interaction paradigms

7.3.2 Physical Substitution, Social Implication, and the Need for Secure Haptics

Long distance communication has become very popular with the advances in the Internet and its applications. With utility applications such as Skype, the world has diminished in size, and video conferencing is a breeze. Seeing and hearing our loved ones does not require traveling anymore since all we need is a laptop and a high speed connection. Nevertheless, people do make an effort to be with the important people in their life because video is not a substitution to physical bonding. With the introduction of haptic multimedia applications, we can engage our touch modality in a tele-conferencing system. Through the advancement of haptic technology, we might feel a lesser need to be physically close to other people in the same geographic location. Hence, our dependence on future technology might change our physical requirements. This, in turn, could lead to unforeseen social behaviors and complications. However, touch is essential for the development of healthy social and psychological behaviors. For instance, infants who do not receive affectionate touch after birth can die within days or develop behaviors similar to autism [46]. Other research has found that touching a loved one can dramatically reduce pain/stress during a medical procedure [61]. Humans rely on the grounding effect of touching people, not devices! For example, the social acceptance of computer-based interpersonal communication (such as hugging a robotic device that represents a remote person) will affect the deployment of haptic interfaces. At the same time, these HAVE applications impose several security and privacy concerns, particularly when used for interpersonal communication like between a parent and child or between remote lovers. Another issue is the susceptibility of communication channels; a malicious party might compromise and obtain access to the haptic information/device. Would they be able to illegally touch/interact with legitimate users? Questions such as this remain research issues, despite the few existing efforts [16].

7.4 Machine Haptics

Although early prototypes of haptic devices were implemented several decades ago, it was not until the early 1990s, with the inauguration of the PHANToM devices, that development of such technologies took off. Nonetheless, haptic technologies still suffer from several challenges that limit their large-scale deployment in a wider spectrum of applications. The following section summarizes some of the outstanding issues that the haptic research community is currently tackling.

7.4.1 Novel Sensing and Actuation Methods

Three types of actuators are commonly employed for kinesthetic haptic devices: electrical, hydraulic, and pneumatic actuators. Each of these suffers from weaknesses while possessing some strengths. For instance, electrical actuators have a relatively low bandwidth and produce small torques with respect to their size. The other two types of devices (hydraulic and pneumatic) are disadvantageously larger in size, so their employment undoubtedly results in bulkier haptic devices. Pneumatic actuators suffer from stiffness and low bandwidth, which makes them impractical for applications that require dexterity (e.g., simulated surgery). Hydraulic actuators have high bandwidth, but they are complex, pricy, and relatively unsafe. There exist several constraints, such as cost, size, weight, robustness, controllability, and bandwidth, that are impeding the spread of haptic devices. For the haptic devices to become popular, novel sensing and actuation methods are yet to be uncovered. Such technologies will be characterized by low manufacturing and operation cost, small size, light weight, high fidelity, energy efficiency, and high mobility. We expect that such methods will be revealed in the near future, particularly for tactile interfaces.

Natural haptic interaction with the environment occurs through multiple points of interaction. For instance, when we grasp an object with our fingers, each finger exerts a force on that object. Today, numerous haptic devices support only one point of interaction. This means that the exchange of forces between the user and the remote or virtual object occurs through only one point of contact. The support of multiple points of interaction is impeded by a haptic device's bulkiness. Multiple points of interaction will require more actuating mechanisms, which will increase the size and complexity of the device, especially for kinesthetic devices. Obviously, such devices are put into motion by actuators. Unfortunately, there is a trade-off between quality of actuation and bulkiness: the higher the quality of the actuators, the bulkier they are. For instance, to increase the dexterity of a device, more degrees of freedom are required, implying that more actuators should be mounted on the device. Moreover, to amplify the realism of the application, multiple points of interaction must be allowed. This will dramatically increase the complexity of the hardware as well as the software. Therefore, to optimize performance for a specific set of applications, device designers must take the above-mentioned realities into account in order to reach a compromise between device size and experiential realism.

7.4.2 Hardware Design and Cost

Most force feedback devices consume more power compared to audiovisual devices because they generate physical energy. In addition, generating physical energy requires mechanisms that are much heavier. Eventually, these devices are characterized by poor power to weight ratios, so although they could still be used in a desktop environment, it would be challenging to make them portable. There have been endeavors to make them lightweight by using strings, but the bulkiness of mechanisms and actuators and the issue of power consumption are still obstacles for mobile environments. Furthermore, due to the fact that force feedback devices generate mechanical energy, an ergonomic design to evade unnecessary movement when moving around the force feedback device should be taken into account. Tactile devices using vibrotactile actuators are lightweight, do not consume much energy compared to other actuators (pneumatic or piezo-electric actuators for example), and are used in most hand-held devices. However, in order to provide rich feedback, the tactile devices need to be equipped with many tactile actuators and cover some part of human body. Furthermore, since the actuators should be kept in physical contact with the human body, users need to wear the device or attach it on the skin. Many kinds of wearable tactile devices have been developed, but comfort has been an issue.

Most commercially available haptic devices are too expensive for personal use, which limits the deployment of HAVE applications significantly. For instance, applications that simulate minimally invasive surgeries can cost more than a million dollars. While such applications can be afforded by big institutions such as universities and hospitals, other devices that have been fashioned for personal use, while costing much less, still struggle to find a receptive market. For instance, an at-home haptic rehabilitation application with a multi-thousand dollar price tag is unlikely to be purchased by patients, regardless of the benefits.

7.4.3   Wireless Portable Haptic Devices

Most haptic devices are stationary (desk-grounded), so the user's workspace is limited (one meter spherical radius in the case of CyberGrasp[1]). As a result, the haptic application scope is limited. Walking-based rehabilitation and immersive virtual environments are two examples of haptic applications where free movement is required. To overcome this limitation, user-grounded portable devices are being introduced. There are two types of such devices: (1) palm or forearm grounded devices and (2) back plate grounded arm exoskeletons. Both types still need wired connections to a computer to transfer haptic data. Therefore, a wireless connection between portable haptic devices and computers is an essential step toward an intuitive, transparent haptic interface. The high transmission rate of haptic data will be one of the main constraints in the design of wireless haptic devices (around 1 kHz). However, the advancement of haptic data compression techniques will make this constraint easier to overcome. Therefore, depending on the data rate of the compressed haptic data, a wireless technology will be selected. Another constraint is the extra weight of the portable power supply of the portable devices, which should be optimized to add as little extra weight as possible to the haptic devices. Otherwise, users experience fatigue during lengthy simulations.

---

[1]http://www.immersion.com.

Mobile devices represent a significant challenge to human–computer interaction designers. Their small physical stature implies that traditional user interface technologies (such as keyboards, mice, graphic displays) are not as efficient as they are with regular computers. Furthermore, sound is of limited value as an interaction means for mobile devices due to ambient noises and/or distractions caused by other people in the nearby surroundings. Therefore, we believe that touch is going to become a lot more popular in the mobile industry (particularly tactile feedback). Indeed, several handsets today are already using tactile feedback for tasks such as pressing a virtual key on a touch screen, identifying information on a web page, message notification alerts, interpersonal communication, gaming and entertainment, and helping the visually impaired/blind. Examples of such existing devices include the LG VX10000 (Verizon Voyager) device and the TouchSense Tactile Feedback system by Immersion.

7.5 Computer Haptics

Although significant progress has been made in the computer haptic domain, particularly in the fields of haptic modeling, collision detection, and force computation, several issues remain unresolved. Here are few of these challenges:

7.5.1 Haptics on Chip

With the increasing complexity of virtual environments, including deformable objects and the demand for more precise force and tactile calculation, the computationally intensive haptic rendering algorithms are becoming more complex and time consuming. This could become a bottleneck for multi-modal applications. The same challenge had been affecting the computer graphics area and eventually led to the popularity of Graphic Processing Units (GPUs). Recently, some complex 6-DOF haptic rendering algorithms with deformable bodies have been successfully implemented using a GPU, but the computation resource is shared with the graphics pipeline [25]. In order to perform independent calculations, a specific haptics card or Haptic Processing Unit (HPU) could be a breakthrough for computationally intensive haptic rendering.

7.5.2 Accessibility and Popularity: Haptic Plug-Ins

One trend in haptics is the increasing availability and popularity of open source and general-purpose haptic rendering libraries and APIs, which provide implementations for core rendering algorithms that can be re-used in a wide spectrum of haptic applications. With the development of web technology, audio visual multimedia became easily accessible through the Internet, and web browsers became a major tool for entertainment, communication, social networking, etc. In order to penetrate the consumer market, HAVE applications also need to be accessible through web browsers. Therefore, plug-ins should be developed for different browsers to enable the distribution and rendering of haptic media using standard formats that are independent of haptic devices. Currently, WebGL, a 3D graphics API implemented in a web browser without plug-ins, is under development and will make interactive 3D content easily available on the Internet.

7.5.3 Reality-Based Modeling: Haptic Rendering Fidelity

Haptic rendering algorithms are increasingly oriented toward representing and generating realistic interactions by imitating physical world interactions. Fidelity and realism of haptic interaction are crucial, particularly for simulation and training applications that are intended to convey mechanical skills that will eventually be applied and experienced in the real world.

A key factor for the success of haptic rendering algorithms is the modeling of physical properties of simulated objects. To accomplish this goal, two approaches are utilized when developing haptic models. The first approach is to collect and read physical properties for an existing object using techniques such as haptic scanning or using a pen-like device that reads haptic interaction data. The second approach is the design of algorithms that generate physics-based haptic models to represent and simulate complex physical properties. Additionally, the ability to model deformation properties is a major contribution to high fidelity of haptic interactions.

7.5.4 Contact Stability and Transparency

It has always been a challenge to render realistic contact forces while retaining the stable behavior of human–environment contact. This becomes a bigger challenge when considering networked haptic applications for several reasons. First, the master and slave parties are energetically coupled and dependent on the network performance. Second, several network parameters have severe impacts on the stability of haptic rendering, including time delay, delay jitter, and packet loss. It is a well-established fact that transparency and stability are both necessary but contradicting requirements for any haptic device. This is especially prominent in networked haptic applications where the trade-off between such requirements is more dramatic. It is also well known that network delay and jitter have a negative impact on device stability. In order to compensate for such effects and maintain an acceptable measure of stability, application designers are obliged to reduce the transparency of the experience. Such unpleasant trade-offs can dramatically reduce the quality of experience for networked applications, especially in applications like tele-surgery, where a high degree of realism is required. Clever control schemes are needed to obtain the best possible results in such cases.

7.6 Multimedia Haptics

A key requirement for future HAVE applications is to accommodate future interaction paradigms such as smell, motion-based user input, biometric-based interactions, and brain–computer interactions. A trend in haptic technologies is the design and development of novel adaptive interfaces that adjust themselves to the user's needs and context. Several other technologies are correlated with the development of adaptive interfaces, such as affective computing (to measure the user's emotional status) and ambient intelligence (smart environments that proactively adapt to a user's preferences and needs). Haptic modality could be the next wave in the development of ambient intelligent systems and applications. A potential trend in haptic technologies is the investigation of how all these future interaction approaches can be integrated into haptics applications.

7.6.1 The Need for Standard Representation

Even with the many efforts made to standardize haptic data representation, such as HAML and the ongoing MPEG-V framework, no universally accepted standards for haptic interaction and data representation exist. Making haptic device plug-and-play interfaces will remain far from reality until such a standard becomes commonplace. The need for a standard haptics data representation is highly relevant.

7.6.2 Consistency and Synchronization

When multiple users are interacting within a shared haptic environment, not only should the interaction forces be distributed to the different network users, but they should also be kept synchronized. Maintaining consistency can be defined as the process of synchronizing a haptic and/or virtual scene among all networked users. This process is a key factor in providing consistent and compelling feedback for all the application participants. Several factors dramatically contribute to consistency, including user actions, the virtual scene, network conditions, and the number of users.

7.6.3 Scalability

The presence of more than two users in a C-HAVE application may be important for applications such as gaming environments, where large numbers of players should be supported. Currently, the possibility of having a large number of networked users interacting with a shared haptic environment is still far from reality. Fortunately, cooperative tasks do not usually require more than five users. Furthermore, it is desirable for a collaboration to have a certain number of users by means of different haptic devices. Other scalability issues include the limitation of the number of objects and the complex of the object simulation that populates the environment.

7.6.4 Network Performance

Network conditions such as delay, jitter, loss of packets, out-of-order delivery, duplicity, and bandwidth have severe impacts on the feasibility of tele-haptic systems. It interferes with our sense of touch, which expects instantaneous feedback and information. Additionally, the touch sense is far more sensitive in terms of responsiveness than vision or hearing. The amount of 30–60 frames per second is needed to visually display a believable constant motion; however, human touch requires updates 1,000 times per second to provide a realistic touch feeling. This update rate is why haptic interaction is very sensitive to network performance. We envision research that focuses on, among other things, the design, evaluation, and analysis of haptic data reduction techniques in order to improve packet transmission in haptic-enabled tele-operation systems, and to reduce the dimensionality in the inherently massive haptic datasets. The discovery of HAVE data reduction methods within the context of haptic data mining and knowledge will primarily help facilitate the analysis of the inherently high-dimensional haptic datasets.

7.6.5 Haptic Memories

Almost a century ago, many memories of a person were virtually lost as soon as the remembered person left the space and time of their loved ones . Only a few decades later, thanks to advances in audio and visual research and development, recording such memories finally became possible. Through the advances in photographic image technology and audio recordings, and much later, video cameras, and DVDs, we gained the ability to see into the past much better than our biological memories would allow. These media have facilitated the recovery of memories, and many of us now sometimes sit around the old picture album to reflect and remember the nice memories we have of our loved ones. Would it not be exciting to be able to restore the smell, touch and hug of remembered ones whenever we desired their affection?

Would it not be interesting to share our parents' affections for us with our children, along with their picture and audio recordings, even after they have passed away? Would it not be interesting to recall past physical affection memories from our childhood? These ideas have triggered a new direction in research concerned with recording, storing, retrieving, and playing back haptic stimuli to bring our memories with far or lost loved ones; this is what we call haptic memories!

7.6.6 Mobile Haptics

Discrete information is like feeling the vibrations of a mobile phone when you receive a message. You have to pick up the phone and to look at it if you want to know whether you missed an SMS or a call. Just when you pick up the phone (the motion can be detected by an accelerometer sensor) the phone could vibrate in a special way (rhythm) so that you immediately know who called you, or even better, it could change its surface somehow. There could also be a way to read the message (similar but not exactly the same as Braille or Morse code) by only touching the phone. In this way, you could receive hidden information during a business meeting. How many times do you see people walking around looking at their phone? Would not it be nice to receive the information without looking at the phone, and thus have the ability to focus on something else? For instance, you want to use your mobile phone to navigate and to make a call at the same time. The navigation part could be done by "feeling" the direction. However, it still remains a challenge to find intuitive ways to transmit and render such information.

7.6.7 Haptics and Security

Today, 70% of the work done by the biometrics market and the industry leaders is dedicated to fingerprints, face recognition patterns, and hand geometry, leaving the remaining (30%) of work dedicated to voice, iris, and signature recognition, middleware, and multi-biometrics fields. Furthermore, biometric technology is playing an important role in developing systems that combine different parameters, such as intrusiveness, cost, distinctiveness, and effort, in the same package. At the present time, fingerprint technology can be found in notebook computers to authenticate individuals based on the physical attributes of their fingertips. However, computer technology, such as faster processors, advanced graphics cards, and multimedia systems, are becoming more affordable with the rapid advancement of the technological revolution. With this technology, the above-mentioned daily life environments can be simulated as computer-generated imagery.

As we have seen, haptic technology is growing in the disciplines linked with human–computer interaction. Haptics is applicable across nearly all areas of computing, and it interfaces between human and computer. Thus, haptics can be used to capture the human-haptic system movements performed during a particular task interaction. Those human psychomotor movements can be used to categorize a behavioral path that can be used for verification and/or authentication purposes. Haptic devices enable the characterization of personalized user-specific physical and biological parameters. When involved in a C-HAVE application, parameters such as forces, end-effector positions, torque, and velocity across the different axes can all be obtained. With the measurements of these parameters, it is possible to identify a user with a rigorous level of precision. The quantification and measurement methodology for such parameters can be a suitable and compelling mechanism to be implemented in a biometric system. Applications of such a system are vast and range from national security applications to access control.

In addition, the protection of haptic information through digital watermarking needs to be deeply investigated. There is a need to examine the role of multisensory feedback in the perception of a watermark embedded in a HAVE application. In particular, the following questions need to be addressed: Do watermarks inspected using multimodal feedback (haptic, audio, and visual) result in very different detection thresholds from those detected using a single sensory modality (touch-only, audio-only, or vision-only)? Also, does visual feedback, when presented together with haptic feedback, improve the perception of watermarks embedded in 3D environments?

7.7 Closing Remarks

It is true that these days haptics appear everywhere; gaming and phones are universally spread through environments. However, a lot remains to be done before the incredible loop formed by our sense of touch and our brain is efficiently assisted by computer applications. Presently, human dexterity (manipulation) assisted by computers appears only in a select few areas. There are challenges in new haptic interfaces and the application of haptic technology. In our opinion, focus should initially lie in the applicability of haptic technologies, from which new hardware and software requirements will naturally become evident and drive the haptic revolution.

References

1. Force dimension (2011). URL http://www.forcedimension.com/, accessed on August 8th, 2011
2. Hapi haptics engine (2011). HAPI Haptics Engine:http://www.ohloh.net/p/HAPI, accessed on August 8th, 2011
3. Logitech (2011). URL http://www.logitech.com, accessed on August 8th, 2011
4. Qhull library (2011). URL http://www.qhull.org, accessed on August 8th, 2011
5. Quanser (2011). URL http://www.quanser.com, accessed on August 8th, 2011
6. Reachin technologies ab (2011). URL http://www.reachin.se, accessed on August 8th, 2011
7. The rutgers haven - a haptic, auditory, and visual environment (2011). URL http://www.cs.rutgers.edu/~dpai/mcl/haven.html, accessed on August 8th, 2011
8. Sensable inc. (2011). URL http://www.sensable.com/index.htm, accessed on August 8th, 2011
9. Abolmaesumi, P., Hashtrudi-Zaad, K., Thompson, D., Tahmasebi, A.: A haptic-based system for medical image examination. In: Proceedings of IEEE 26th Annual International Conference of the Engineering in Medicine and Biology Society, vol. 3, pp. 1853–1856. San Francisco, CA (2004)
10. Adams, R., Hannaford, B.: Stable haptic interaction with virtual environments. IEEE Trans. Rob. Autom. **15**(3), 465–474 (1999)
11. Adams, R., Hannaford, B.: Control law design for haptic interfaces to virtual reality. IEEE Trans. Control. Syst. Technol. **10**(1), 3–13 (2002)
12. Al Osman, H., Eid, M., Iglesias, R., El Saddik, A.: Alphan: Application layer protocol for haptic networking. In: IEEE International Workshop on Haptic, Audio and Visual Environments and Games, 2007. (HAVE 2007) pp. 96–101 (2007)
13. Alamri, A., Cha, J., El Saddik, A.: Ar-rehab: An augmented reality framework for poststroke-patient rehabilitation. IEEE Trans. Instrum. Meas. **59**(10), 2554–2563 (2010)
14. Alatan, A.A., Yemez, Y., Güdükbay, U., Zabulis, X., Müller, K., Ç. E. Erdem, Weigel, C., Smolic, A.: Scene representation technologies for 3DTV—A survey. IEEE Trans. Circ. Syst. Video Tech. **17**(11), 1587–1605 (2007)
15. Alles, D.S.: Information transmission by phantom sensations. IEEE Trans. Man Mach. Syst. **MMS-11**(1), 85–91 (1970)
16. Alsulaiman, F., Cha, J., El Saddik, A.: User identification based on handwritten signatures with haptic information. In: Proceedings of the EuroHaptics 2008 Conference. Madrid, Spain (2008)
17. Ambrosi, G., Bicchi, A., De Rossi, D., Scilingo, E.: The role of contact area spread rate in haptic discrimination of softness. In: Proceedings of the 1999 IEEE International Conference on Robotics and Automation, 1999. vol. 1, pp. 305–310 (1999)

18. Andrews, S., Mora, J., Lang, J., Lee, W.: Hapticast: A physically-based 3d game with haptic feedback. In: Proceedings of Future Play 2006. London, ON, Canada (2006)
19. Angeles, J.: Fundamentals of Robotic Mechanical Systems: Theory, Methods, and Algorithms. Springer, New York, 2003 (1943)
20. Avila, R.S., Sobierajski, L.M.: A haptic interaction method for volume visualization. In: Proceedings of IEEE Visualization Conference, pp. 197–204 (1996)
21. Avizzano, C., Marcheschi, S., Angerilli, M., Fontana, M., Bergamasco, M., Gutierrez, T., Mannegeis, M.: A multi-finger haptic interface for visually impaired people. In: The 12th IEEE International Workshop on Robot and Human Interactive Communication, pp. 165–170 (2003)
22. Baheux, K., Yoshizawa, M., Tanaka, A., Seki, K., Handa, Y.: Diagnosis and rehabilitation of patients with hemispatial neglect using virtual reality technology. In: Proceedings of the 26th Annual International Conference of the IEEE EMBS (2004)
23. Bailenson, J., Yee, N., Merget, D., Schroeder, R.: The effect of behavioral realism and form realism of real-time avatar faces on verbal disclosure, nonverbal disclosure, emotion recognition, and copresence in dyadic interaction. Presence 15(4), 359–372 (2006)
24. Barbagli, F., Salisbry K., J., Devengenzo, R.: Enabling multi-finger, multi-hand virtualized grasping. In: Proceedings of ICRA '03, IEEE International Conference on Robotics and Automation, 2003. vol. 1, pp. 809–815 (2003)
25. Barbič, J., James, D.: Time-critical distributed contact for 6-dof haptic rendering of adaptively sampled reduced deformable models. In: Proceedings of the 2007 ACM SIG-GRAPH/Eurographics Symposium on Computer Animation, SCA '07, Eurographics Association, pp. 171–180. Aire-la-Ville, Switzerland, (2007)
26. Basdogan, C., De, S., Kim, J., Muniyandi, M., Kim, H., Srinivasan, M.: Haptics in minimally invasive surgical simulation and training. IEEE Comput. Graph. Appl. 24(2), 56–64 (2004)
27. Basdogan, C., Ho, C., Srinivasan, M.: Virtual environments for medical training: Graphical and haptic simulation of laparoscopic common bile duct exploration. IEEE/ASME Trans. Mechatron. 6, 269–285 (2001)
28. Basdogan, C., Ho, C., Srinivasan, M.A.: A ray-based haptic rendering technique for displaying shape and texture of 3d objects in virtual environments. In: Winter Annual Meeting of ASME'97, pp. 77–84 (1997)
29. Basdogan, C., Ho, C.H., Srinivasan, M.A., Slater, M.: An experimental study on the role of touch in shared virtual environments. ACM Trans. Comput. Hum. Interact. 7, 443–460 (2000)
30. Baxter, W., Scheib, V., Lin, M., Manocha, D.: Dab: Interactive haptic painting with 3d virtual brushes. In: SIGGRAPH' 01, pp. 461–468 (2001)
31. Becker, J., Thakor, N., Gruben, K.: A study of human hand tendon kinematics with applications to robot hand design. In: Proceedings of the 1986 IEEE International Conference on Robotics and Automation. vol. 3, pp. 1540–1545 (1986)
32. Békésy, G.V.: Funneling in the nervous system and its role in loudness and sensation intensity on the skin. J. Acoust. Soc. Am. 30(5), 399–412 (1958)
33. Bergamasco, M., Alessi, A.A., Calcara, M.: Thermal feedback in virtual environments. Presence 6, 617–629 (1997)
34. Bergamasco, M., Allotta, B., Bosio, L., Ferretti, L., Parrini, G., Prisco, G., Salsedo, F., Sartini, G.: An arm exoskeleton system for teleoperation and virtual environments applications. In: Proceedings of the 1994 IEEE International Conference on Robotics and Automation, 1994. vol. 2, pp. 1449–1454 (1994)
35. Bergamasco, M., Frisoli, A., Barbagli, F.: Haptics technologies and cultural heritage applications. In: Proceedings of the Computer Animation (CA), pp. 25–32 (2002)
36. Berkelman, P., Hollis, R., Baraff, D.: Interaction with a real time dynamic environment simulation using a magnetic levitation haptic interface device. In: Proceedings of the 1999 IEEE International Conference on Robotics and Automation, 1999. vol. 4, pp. 3261–3266 (1999)
37. Bernstein, N., Lawrence, D., Pao, L.: Friction modeling and compensation for haptic interfaces. In: Eurohaptics Conference, 2005 and Symposium on Haptic Interfaces for Virtual Environment and Teleoperator Systems, 2005. World Haptics 2005. First Joint, pp. 290–295 (2005)

38. Bethea, B.T., Okamura, A.M., Kitagawa, M., Fitton, T.P., Cattaneo, S.M., Gott, V.L., Baumgartner, W.A., Yuh, D.D.: Application of haptic feedback to robotic surgery. J. Laparoendosc. Adv. Surg. Tech. **14**(3), 191–195 (2004)
39. Bhasin, Y., Liu, A., Bowyer, M.: Simulating surgical incisions without polygon subdivision. Medicine Meets Virtual Reality, vol. 111, pp. 43–49. IOS Press, Amsterdam, The Netherlands (2005)
40. Blanch, R., Ferley, E., Cani, M., Gascuel, J.: Non-realistic haptic feedback for virtual sculpture. Technical Report No RR-5090. INRIA, U.R. Rhone-Alpes (2004)
41. Boff, K., Kaufman, L., Thomas, J. (eds.): Psychophysical measurement and theory, Handbook of Perception and Human Performance. Wiley, New York (1986)
42. Boian, R., Deutsch, J., Lee, C.S., Burdea, G., Lewis, J.: Haptic effects for virtual reality-based post-stroke rehabilitation. In: Proceedings of the 11th Symposium on Haptic Interfaces for Virtual Environment and Teleoperator Systems, 2003. (HAPTICS 2003). pp. 247–253 (2003)
43. Boie, R.: Capacitive impedance readout tactile image sensor. In: Proceedings of the 1984 IEEE International Conference on Robotics and Automation. vol. 1, pp. 370–378 (1984)
44. Bolanowski, S.J., Gescheider, G.A., Verrillo, R.T., Checkosky, C.M.: Four channels mediate the mechanical aspects of touch. J. Acoust. Soc. Am. **84**(5), 1680–1694 (1988)
45. Bonanni, L., Vaucelle, C., Lieberman, J., Zuckerman, O.: Taptap: A haptic wearable for asynchronous distributed touch therapy. In: Proceedings of ACM CHI, pp. 580–585 (2006)
46. Bonnani, L., Vaucelle, C.: A framework for haptic psycho-therapy. In: IEEE International Conference of Pervasive Services (2006)
47. Boukerche, A., Shirmohammadi, S., Hossain, A.: Moderating simulation lag in haptic virtual environments. In: ANSS '06: Proceedings of the 39th annual Symposium on Simulation, IEEE Computer Society, pp. 269–277. Washington, DC, USA (2006)
48. Bouzit, M., Burdea, G., Popescu, G., Boian, R.: The rutgers master ii-new design force-feedback glove. ASME Trans Mechatron **7**, 256–263 (2002)
49. Bouzit, M., Chaibi, A., DeLaurentis, K., Mavroidis, C.: Tactile feedback navigation handle for the visually impaired. In: Proceedings of IMECE2004, 2004 ASME International Mechanical Engineering Congress and RD&D Expo. Anaheim, California USA (2004)
50. Brave, S., Daley, A.: Intouch: A medium for haptic interpersonal communication. In: Proceedings of ACM CHI'97, pp. 363–364 (1997)
51. Brewster, S.: The impact of haptic "touching" technology on cultural applications. In: Proceedings of EVA, vol. 28, pp. 1–14. Vasari, UK (2001)
52. Broeren, J., Georgsson, M., Rydmark, M., Sunnerhagen, K.: Virtual reality in stroke rehabilitation with the assistance of haptics and telemedicine? In: Proceedings of 4th International Conference on Disability, Virtual Reality & Associated Technologies. Veszprem, Hungary (2002)
53. Brooks Jr., F.P., Ouh-Young, M., Batter, J.J., Jerome Kilpatrick, P.: Project grope - haptic displays for scientific visualization. In: SIGGRAPH '90: Proceedings of the 17th annual conference on Computer graphics and interactive techniques, vol. 24, pp. 177–185. ACM, New York, USA (1990)
54. Brooks, T.: Telerobot response requirements. Tech. rep., Lanham, Md.: STX Robotics (1990)
55. Brown, J.M., Colgate, J.E.: Passive implementation of multibody simulations for haptic display. In: Proceedings of the 1997 ASME International Mechanical Engineering Congress and Exhibition, vol. DSC-61, pp. 85–92 (1997)
56. Burdea, G., Zhuang, J., Roskos, E., Silver, D., Langrana, N.: A portable dextrous master with force feedback. Presence Teleoperator Virtual Environ. **1**, 18–28 (1992)
57. Burdea, G.C., Coiffet, P.: Virtual Reality Technology, 2nd edn. Wiley, Hoboken, New jersey (2003)
58. Buttolo, P., Oboe, R., Hannaford, B.: Architectures for shared haptic virtual environments. Comput. Graph. **21**(4), 421–429 (1997) (Haptic Displays in Virtual Environments and Computer Graphics in Korea)
59. Canarie: Canada's advanced research and innovation network. http://canarie.ca/en/home (2011), accessed on August 8th, 2011

60. Cardin, S., Thalmann, D., Vexo, F.: A wearable system for mobility improvement of visually impaired people. Vis. Comput. J. **23**(2), 109–118 (2006)
61. Carey, B.: Holding loved one's hand can calm jittery neurons. The New York Times (2006)
62. Carignan, C., Cleary, K.: Closed-loop force control for haptic simulation of virtual environments. Haptic-e **1**(2), 1–14 (2000)
63. Cascio, J., Sathian, K.: Temporal cues contribute to tactile perception of roughness. J. Neurosc. **21**(14), 5289–5296 (2001)
64. Caselli, S., Magnanini, C., Zanichelli, F., Caraffi, E.: Efficient exploration and recognition of convex objects based on haptic perception. In: Proceedings of the 1996 IEEE International Conference on Robotics and Automation, 1996. vol. 4, pp. 3508–3513 (1996)
65. Cen, Z., Mutka, M., Liu, Y., Goradia, A., Xi, N.: Qos management of supermedia enhanced teleoperation via overlay networks. In: 2005 IEEE/RSJ International Conference on Intelligent Robots and Systems, 2005. (IROS 2005). pp. 1630–1635 (2005)
66. Cha, J., Eid, M., Barghout, A., Rahman, A.M., El Saddik, A.: Hugme: Synchronous haptic teleconferencing. In: Proceedings of the 17th ACM international conference on Multimedia, MM '09, pp. 1135–1136. ACM, New York, NY, USA (2009)
67. Cha, J., Eid, M., El Saddik, A.: Dibhr: Depth image-based haptic rendering. In: Ferre, M. (ed.) Haptics: Perception, Devices and Scenarios, Lecture Notes in Computer Science, vol. 5024, pp. 640–650. Springer, Berlin (2008)
68. Cha, J., Eid, M., El Saddik, A.: Touchable 3d video system. ACM Trans. Multimed. Comput. Comm. Appl. **5**(4), 29:1–29:25 (2009)
69. Cha, J., Ho, Y., Kim, Y., Ryu, J., Oakley, I.: A framework for haptic broadcasting. IEEE Multimed. **16**(3), 16–27 (2009)
70. Cha, J., Kim, S., Ho, Y., Ryu, J.: 3D video player system with haptic interaction based on depth image-based representation. IEEE Trans. Consum. Electron. **52**(2), 477–484 (2006)
71. Cha, J., Ryu, J., Kim, S., Ahn, B.: A haptically enhanced broadcasting system. In: Eurohaptics Conference, 2005 and Symposium on Haptic Interfaces for Virtual Environment and Teleoperator Systems, 2005. World Haptics 2005. First Joint, pp. 515–516 (2005)
72. Cha, J., Ryu, J., Kim, S., Eom, S., Ahn, B.: Haptic interaction in realistic multimedia broadcasting. In: 5th Pacific-Rim Conference on Multimedia (PCM), vol. LNCS 3333, pp. 482–490 (2004)
73. Chandra, K.: The Wiley Encyclopedia of Telecommunications. Statistical Multiplexing. Wiley-Interscience, New York (2003)
74. Chen, H., Sun, H.: Real-time haptic sculpting in virtual volume space. In: Proceedings of the ACM Symposium on Virtual Reality Software and Technology, pp. 81–88. Hong Kong, China (2002)
75. Cheong, J., Niculescu, S.I., Annaswamy, A., Srinivasan, M.: Motion synchronization in virtual environments with shared haptics and large time delays. In: Eurohaptics Conference, 2005 and Symposium on Haptic Interfaces for Virtual Environment and Teleoperator Systems, 2005. World Haptics 2005. First Joint, pp. 277–282 (2005)
76. Chial, V., Greenish, S., Okamura, A.: On the display of haptic recordings for cutting biological tissues. In: Haptic Interfaces for Virtual Environment and Teleoperator Systems, 2002. HAPTICS 2002. Proceedings. 10th Symposium, pp. 80–87 (2002)
77. Choi, S., Tan, H.: Aliveness: Perceived instability from a passive haptic texture rendering system. In: Intelligent Robots and Systems, 2003. (IROS 2003). Proceedings. 2003 IEEE/RSJ International Conference on, vol. 3, pp. 2678–2683 (2003)
78. Cohen, A., Chen, E.: Six degree-of-freedom haptic system as a desktop virtual prototyping system. In: Proceedings of the First International Workshop on Virtual Reality and Prototyping, vol. 1, pp. 97–106 (1999)
79. Cohen, J.D., Lin, M.C., Manocha, D., Ponamgi, M.: I-collide: An interactive and exact collision detection system for large-scale environments. In: Proceedings of the 1995 Symposium on Interactive 3D graphics, I3D '95, pp. 189–196. ACM, New York, NY, USA (1995)

80. Colgate, J., Brown, J.: Factors affecting the z-width of a haptic display. In: Proceedings of the 1994 IEEE International Conference on Robotics and Automation, 1994. vol. 4, pp. 3205–3210 (1994)
81. Colgate, J., Grafing, P., Stanley, M., Schenkel, G.: Implementation of stiff virtual walls in force-reflecting interfaces. In: Virtual Reality Annual International Symposium, 1993.,1993 IEEE, pp. 202–208 (1993)
82. Colgate, J., Peshkin, M., Wannasuphoprasit, W.: Nonholonomic haptic display. In: Proceedings of the 1996 IEEE International Conference on Robotics and Automation, 1996. vol. 1, pp. 539–544 (1996)
83. Colgate, J.E., Schenkel, G.G.: Passivity of a class of sampled-data systems: Application to haptic interfaces. J. Robot. Syst. **14**(1), 37–47 (1997)
84. Connor, B., Wing, A., Humphreys, G., Bracewell, R., Harvey, D.: Errorless learning using haptic guidance: Research in cognitive rehabilitation following stroke. In: 4th Internatinal Conference on Disability, Virtual Reality and Associated Technology, pp. 77–84. Veszprem, Hungary (2002)
85. Conti, F., Barbagli, F., Morris, D., Sewell, C.: Chai 3d – an open-source library for the rapid development of haptic scenes. In: IEEE World Haptics, pp. 140–145 vol.3. Pisa, Italy (2005)
86. Corsini, M., Drelie Gelasca, E., Ebrahimi, T.: A Multi-Scale Roughness Metric for 3D Watermarking Quality Assessment. In: Workshop on Image Analysis for Multimedia Interactive Services 2005, April 13–15, Montreux, Switzerland., ISCAS. SPIE (2005)
87. Costa, M.A., Cutkosky, M.R.: Roughness perception of haptically displayed fractal surfaces. In: ASME Dynamic Systems and Control Division, vol. 69, pp. 1073–1079 (2000)
88. CuteCircuit: Wearable technology. http://www.cutecircuit.com/products/wearables/ (2008). URL http://www.cutecircuit.com/products/wearables/, accessed on August 8th, 2011
89. Dachille, F., Qin, H., Kaufman, A., El-Sana, J.: Haptic sculpting of dynamic surfaces. In: Proceedings of 1999 Symposium on Interactive 3D Graphics, pp. 103–110. Atlanta, Georgia (1999)
90. Darian-Smith, I.: The sense of touch: Performance and peripheral neural processes. In: Handbook of Physiology, The Nervous System, Sensory Processes, **3**, 739–788 (1984)
91. Deepa, M.: vsmileys: Imaging emotions through vibration patterns. Alternative access: Feelings and games 2005, Department of Computer Sciences University of Tampere, Finland (2005)
92. Dennerlein, J., Yang, M.: Haptic force-feedback devices for the office computer: Performance and musculoskeletal loading issues. Human Factors: J. Hum. Factors Ergon. Soc. **43**(3), 278–286 (2001)
93. Desai, J.P., Tholey, G., Kennedy, C.W.: Haptic feedback system for robot-assisted surgery. In: Performance Metrics for Intelligent Systems Workshop, pp. 202–209. Maryland, USA (2007)
94. Deutsch, J., Latonio, J., Burdea, G., Boian, R.: Rehabilitation of musculoskeletal injuries using the rutgers ankle haptic interface: Three case reports. In: Eurohaptics Conference, pp. 11–16. Birmingham, UK (2001)
95. DiFranco, D., Beauregard, G., Srinivasan, M.: The effect of auditory cues on the haptic perception of stiffness in virtual environments. In: Proceedings of the ASME Dynamic Systems and Control Division, pp. 17–22, Atlanta, Georgia (1997)
96. DiMaio, S., Salcudean, S.: Needle insertion modelling and simulation. In: ICRA 2002, pp. 2098–2105 (2002)
97. Diolaiti, N., Niemeyer, G., Barbagli, F., Salisbury, J.: A criterion for the passivity of haptic devices. In: Proceedings of the 2005 IEEE International Conference on Robotics and Automation, pp. 2452–2457 (2005)
98. Dodeller, S.: Transport Layer Protocols for Haptic Virtual Environments, chap. M.S. thesis. Springer, Heidelberg (2004)
99. Durlach, N.I., Pew, R., Aviles, W., DiZio, P.A., Zeltzer, D.L.: Virtual environment technology for training (vett). Tech. rep., BBN Report No. 7661. (1992)

100. Eagleman, D.: Visual illusions and neurobiology. Nat. Rev. Neurosci. **2**(12), 920–6 (2001)
101. Ehmann, S.A., Lin, M.C.: Accurate and fast proximity queries between polyhedra using convex surface decomposition. In: Computer Graphics Forum, pp. 500–510 (2001)
102. Eid, M., Alamri, A., El Saddik, A.: Mpeg-7 description of haptic applications using haml. In: IEEE International Workshop on Haptic Audio Visual Environments and their Applications, (HAVE 2006). pp. 134–139 (2006)
103. Eid, M., Andrews, S., Alamri, A., El Saddik, A.: Hamlat: A haml-based authoring tool for haptic application development. In: Ferre, M. (eds.) Haptics: Perception, Devices and Scenarios, vol. 5024, pp. 857–866. Springer, Heidelberg (2008)
104. Eid, M., Cha, J., El Saddik, A.: Admux: An adaptive multiplexer for haptic-audio-visual data communication. IEEE Trans. Instrum. Meas. **60**(1), 21–31 (2011)
105. Eid, M., Cha, J., Rahal, L., El Saddik, A.: Hugme: A haptic videoconferencing system for interpersonal communication. In: Proceedings of the International Conference Virtual Environments, Human-Computer Interfaces, and Measurement System (VECIMS), vol. 178, pp. 99–102 (2008)
106. Eid, M., Orozco, M., El Saddik, A.: A guided tour in haptic audio visual environment and applications. J. Adv. Media and Comm. **1**(3), 265–297 (2007)
107. El-Far, N., Nourian, S., Zhou, J., Hamam, A., Shen, X., Georganas, N.: A cataract tele-surgery training application in a hapto-visual collaborative environment running over the canarie photonic network. In: IEEE International Workshop on Haptic Audio Visual Environments and their Applications, pp. 29–32 (2005)
108. El-Far, N., Shen, X., Georganas, N.: Applying unison, a generic framework for hapto-visual application development, to an e-commerce application. In: Proceedings of the IEEE Workshop on Haptic Audio Visual Environments and their Applications. Ottawa, ON, Canada (2004)
109. El Saddik, A.: The potential of haptics technologies. IEEE Instrum. Meas. **10**(1), 10–17 (2007)
110. El Saddik, A., Orozco, M., Asfaw, Y., Shirmohammadi, S., Adler, A.: A novel biometric system for identification and verification of haptic users. IEEE Trans. Instrum. Meas. **56**(3), 895–906 (2007)
111. Emura, S., Tachi, S.: Active haptic perception and control of its fixation. In: IEEE International Conference on Robotics and Automation, vol. 2, pp. 976–983 (1999)
112. Esen, H., Yano, K., Buss, M.: A virtual environment medical training system for bone drilling with 3 dof force feedback. In: Proceedings of IEEEIRSJ International Conference on Intelligent Robots and Systems (2004)
113. Fasse, E.D., Hogan, N.: Quantitative measurement of haptic perception. In: ICRA, vol. 4, pp. 3199–3204 (1994)
114. Faust, M., Yoo, Y.: Haptic feedback in pervasive games. In: Third International Workshop on Pervasive Gaming Applications. Dublin, Ireland (2006)
115. Fayad, M., Schmidt, D.C.: Object-oriented application frameworks. Commun. ACM. **40**, 32–38 (1997)
116. Fearing, R.S.: Tactile sensing mechanisms. Int. J. Rob. Res. **9**, 3–23 (1990)
117. Feygin, D., Keehner, M., Tendick, R.: Haptic guidance: Experimental evaluation of a haptic training method for a perceptual motor skill. In: Haptic Interfaces for Virtual Environment and Teleoperator Systems, 2002. HAPTICS 2002. Proceedings. 10th Symposium, pp. 40–47 (2002)
118. Fogg, B., Cutler, L., Arnold, P., Eisback, C.: Handjive: A device for interpersonal haptic entertainment. In: Press, A. (ed.) In Proc CHI'98 Human Factors inComputing Systems, ACM Press, pp. 57–64 (1998)
119. Fourney, D., Carter, J.: Initiating guidance on tactile and haptic interactions. In: Proceedings of Guidelines On Tactile and Haptic Interactions (GOTHI'5) Conference, pp. 1–9 (2005)
120. Frisoli, A., Sotgiu, E., Avizzano, C., Checcacci, D., Bergamasco, M.: Force-based impedance control of a haptic master system for teleoperation. Sensor Review **24**(1), 42–50 (2004)

121. Fukuda, I., Matsumoto, S., Iijima, M., Hikichi, K., Morino, H., Sezaki, K., Yasuda, Y.: A robust system for haptic collaboration over the network. In: Touch in Virtual Environments Conference, pp. 137–157, Prentice Hall, Upper Saddle River, NJ (2002)
122. Gao, Z., Gibson, I.: Haptic B-spline surface sculpting with a shaped tool of implicit surface. Computer-Aided Design Applications 2(1-4), 263–272 (2005)
123. Gautier, L., Diot, C., Kurose, J.: End-to-end transmission control mechanisms for multiparty interactive applications on the internet. In: INFOCOM '99. Eighteenth Annual Joint Conference of the IEEE Computer and Communications Societies. vol. 3, pp. 1470–1479 (1999)
124. Gaven, W.: Everyday listening and auditory icons. Ph.D. thesis, University of California, San Diego (1988)
125. Gaw, D., Morris, D., Salisbury, K.: Haptically annotated movies: Reaching out and touching the silver screen. In: 14th International Symposium on Haptic Interfaces for Virtual Environment and Teleoperator Systems, pp. 287–288 (2006)
126. Gentaz, E., Hatwell, Y.: Geometrical haptic illusions: The role of exploration in the muller-lyer, vertical-horizontal and delboeuf illusions. Psychon. Bull. Rev. 11(1), 31–40 (2004)
127. Gibson, J.J.: The Senses Considered as Perceptual Systems. pp. 1–6, Houghton Mifflin, Boston (1966)
128. Gil, J., Avello, A., Rubio, A., Florez, J.: Stability analysis of a 1 dof haptic interface using the routh-hurwitz criterion. IEEE Trans. Contr. Syst. Tech. 12(4), 583–588 (2004)
129. Gil, J., Sanchez, E.: Control algorithms for haptic interaction and modifying the dynamical behavior of the interface. In: 2nd International Conference on Enactive Interfaces, pp. 17–18, Genoa, Italy (2005)
130. Gillespie, B.: Haptic displays of systems with changing kinematic constraints: The virtual piano action. Ph.D. thesis, Stanford University, Stanford, California (1996)
131. Glencross, M., Marsh, J., Cook, J., Daubrenet, S., Pettifer, S., Hubbold, R.: Divipro: Distributed interactive virtual prototoytping. In: SIGGRAPH'02 ACM Conference, San Antonio, Texas (2002)
132. GmbH, V.: VirtualHand SDK, http://www.cyberglovesystems.com/products/virtual-hand-sdk/overview, accessed on August 8th, 2011
133. Goertz, R.: Fundamentals of general-purpose remote manipulators. Nucleonics 10(11), 36–45 (1952)
134. Goertz, R., Thompson, W.: Electronically controlled manipulator. Nucleonics 12(11), 46–47 (1954)
135. Goldsmith, J., Salmon, J.: Automatic creation of object hierarchies for ray tracing. IEEE Comput. Graph. Appl. 7(5), 14–20 (1987)
136. Gorges, N., Bierbaum, A., Wörn, H., Dillmann, R.: Towards a comprehensive grasping system for armar-iii. In: Human-Centered Robotic Systems (HCRS'06) (2006)
137. Gosline, A.H., Turgay, E., Brouwer, I.: Haptic illusions: What you feel isn't always what you get. In: Proceedings of Human Interface Technologies, pp. 19–22 (2002)
138. Gosselin, F., Bidard, C., Brisset, J.: Design of a high fidelity haptic device for telesurgery. In: Proceedings of the 2005 IEEE International Conference on Robotics and Automation. pp. 206–211, Barcelona, Spain (2005)
139. Green, M., Palfrey, J.: Bright futures: Guidelines for health supervision of infants, children, and adolescents. In: 2nd ed. National Center for Education in Maternal and Child Health. Green, M., Palfrey, J.S., Georgetown University Arlington (2002)
140. Gregory, A., Mascarenhas, A., Ehmann, S., Lin, M., Manocha, D.: Six degree-of-freedom haptic display of polygonal models. In: Proceedings of the conference on Visualization '00, VIS '00, pp. 139–146. IEEE Computer Society Press, Los Alamitos, CA, USA (2000)
141. Guerraz, A., Loscos, C., Widenfeld, H.R.: How to use physical parameters coming from the haptic device itself to enhance the evaluation of haptic benefits in user interface? In: EuroHaptics (2003)
142. Gunther, E., O'Modhrain, S.: Cutaneous grooves: Composing for the sense of touch. J. New Music Res. 32(4), 369–381 (2003)

143. Hackwood, S., beni, G., Hornak, L., Wolfe, R., Nelson, T.J.: A torque-sensitive tactile array for robotics. Int. J. Robotics Res. **2**(2), 46–50 (1983)
144. Hale, K., Stanney, K.: Deriving haptic design guidelines from human physiological, psychophysical, and neurological foundations. IEEE Comput. Graph. Appl. **24**(2), 33–39 (2004)
145. Hamam, A., Eid, M., El Saddik, A., Georganas, N.D.: A quality of experience model for haptic user interfaces. In: Proceedings of the 2008 Ambi-Sys workshop on Haptic user interfaces in ambient media systems, HAS '08, pp. 1:1–1:6. ICST (Institute for Computer Sciences, Social-Informatics and Telecommunications Engineering), ICST, Brussels, Belgium, Belgium (2008)
146. Hamam, A., Nourian, S., El-Far, N., Malric, F., Shen, X., Georganas, N.: A distributed, collaborative and haptic-enabled eye cataract surgery application with a user interface on desktop, stereo desktop and immersive displays. In: IEEE International Workshop on Haptic Audio Visual Environments and their Applications. Ottawa, Canada (2006)
147. Hannaford, B., Ryu, J.H.: Time domain passivity control of haptic interfaces. In: IEEE International Conference on Robotics and Automation, (ICRA 2001), vol. 2, pp. 1863–1869 (2001)
148. Harwin, W.S., Melder, N.: Improved haptic rendering for multi-finger manipulation using friction cone based god-objects. In: EuroHaptics'02 (2002)
149. Hasser, C., Daniels, M.: Tactile feedback with adaptive controller for a force-reflecting haptic display. 2. improvements and evaluation. In: Proceedings of the 1996 Fifteenth Southern Biomedical Engineering Conference, vol. 1, pp. 530–533 (1996)
150. Hayward, V.: Toward a seven axis haptic device. In: IEEE/RSJ International Conference on Intelligent Robots and Systems, pp. 133–139. Pittsburgh, PA, USA (1995)
151. Hayward, V., Armstrong, B.: A new computational model of friction applied to haptic rendering. In: Experimental Robotics VI, Lecture Notes in Control and Information Sciences, vol. 250, pp. 403–412. Springer, Heidelberg (2000)
152. Hayward, V., Astley, O., Cruz-Hernandez, M., Grant, D., Robles-De-La-Torre, G.: Haptic interfaces and devices. Sensor Review **24**(1), 16–29 (2004)
153. Hayward, V., O.R., A.: Performance measures for haptic interfaces. In: 7th International Symposium in Robotics Research, pp. 195–207. Springer (1996)
154. Heller, M.: Visual and tactual texture perception: Intersensory cooperation. Perception Psychophys. **31**, 339–344 (1982)
155. Heller, M.A., Schiff, W.: The Psychology of Touch. Lawrence Erlbaum Associates, pp. 91–114. Hillsdale, NJ (1991)
156. Hespanha, J.P., Mclaughlin, M., Sukhatme, G.S., Akbarian, M., Garg, R., Zhu, W.: Haptic collaboration over the internet. In: Fifth Phantom Users Group Workshop, pp. 9–13. Prentice Hall (2000)
157. Hikichi, K., Morino, H., Fukuda, I., Matsumoto, S., Yasuda, Y., Arimoto, I., Iijima, M., Sezaki, K.: Architecture of haptics communication system for adaptation to network environments. In: IEEE International Conference on Multimedia and Expo. (ICME 2001), pp. 563–566 (2001)
158. Hikichi, K., Morino, H., Yasuda, Y., Arimoto, I., Sezaki, K.: The evaluation of adaptation control for haptics collaboration over the internet. In: IEEE Communications Quality and Reliability (CQR) International Workshop, pp. 218–222 (2002)
159. Hikichi, K., Yasuda, Y., Fukuda, A., Sezaki, K.: The effect of network delay on remote calligraphic teaching with haptic interfaces. In: Proceedings of 5th ACM SIGCOMM workshop on Network and system support for games, NetGames '06. ACM, New York, NY, USA (2006)
160. Hillis, W.D.: A high-resolution imaging touch sensor. Int. J. Robot. Res. **1**(2), 33–44 (1982)
161. Hinterseer, P., Hirche, S., Chaudhuri, S., Steinbach, E., Buss, M.: Perception-based data reduction and transmission of haptic data in telepresence and teleaction systems. IEEE Trans. Signal Process. **56**(2), 588–597 (2008)
162. Hinterseer, P., Steibach, E., Chaudhuri, S.: Model based data compression for 3d virtual haptic teleinteraction. In: International Conference on Consumer Electronics. (ICCE '06). Digest of Technical Papers. pp. 23–24 (2006)

163. Hinterseer, P., Steinbach, E.: A psychophysically motivated compression approach for 3d haptic data. In: 14th Symposium on Haptic Interfaces for Virtual Environment and Teleoperator Systems, pp. 35–41 (2006)

164. Hinterseer, P., Steinbach, E., Hirche, S., Buss, M.: A novel, psychophysically motivated transmission approach for haptic data streams in telepresence and teleaction systems. In: IEEE International Conference on Acoustics, Speech, and Signal Processing, (ICASSP '05). vol. 2, pp. ii/1097–ii/1100 (2005)

165. Hinterseer, R., Steinbach, E., Chaudhuri, S.: Perception-based compression of haptic data streams using kalman filters. In: IEEE International Conference on Acoustics, Speech and Signal Processing, (ICASSP 2006), vol. 5, pp. V–V (2006)

166. Hirche, S., Buss, M.: Transparent data reduction in networked telepresence and teleaction systems. part ii: Time-delayed communication. Presence: Teleoperators and Virtual Environments 16(5), 532–542 (2007)

167. Hirche, S., Hinterseer, P., Steinbach, E., Buss, M.: Transparent data reduction in networked telepresence and teleaction systems. part i: Communication without time delay. Presence: Teleoperators and Virtual Environments 16(5), 523–531 (2007)

168. Ho, C., Basdogan, C., Srinivasan, M.A.: Efficient point-based rendering techniques for haptic display of virtual objects. Presence: Teleoperators and Virtual Environments 8(5), 477–491 (1999)

169. Hollins, M., Risner, S.: Evidence for the duplex theory of tactile texture perception. Percept Psychophys 62, 695–705 (2000)

170. Hollins, M., Seeger, A., Pelli, G., Taylor, R.: Haptic perception of virtual surfaces: Scaling subjective qualities and interstimulus differences. Perception 33(8), 1001–1019 (2004)

171. Hollis, R., Salcudean, S.: Lorentz levitation technology: A new approach to fine motion robotics, teleoperation, haptic interfaces, and vibration isolation. In: Int'l Symposium for Robotics Research. Hidden Valley, PA (1993)

172. Hollis, R., Salcudean, S., Allan, A.: A six-degree-of-freedom magnetically levitated variable compliancefine-motion wrist: Design, modeling, and control. IEEE Trans. Robot, Autom. 7(3), 320–332 (1991)

173. Hollis, R.L.: Butterfly haptics: A high-tech start-up. IEEE Tobotics Autom. Mag. 17, 14–17 (2010)

174. Hossain, S., Rahman, A., El Saddik, A.: Interpersonal haptic communication in second life. In: Haptic Audio-Visual Environments and Games (HAVE), IEEE International Symposium, pp. 1–4 (2010)

175. Howell, J.N., Conatser, R.R., Williams, R.L., Burns, J.M., Eland, D.C.: Palpatory diagnosis training on the virtual haptic back: Performance improvement and user evaluations. Medical Education 108(1), 29–36 (2008)

176. Hubbold, R.J.: Collaborative stretcher carrying: A case study. In: Proceedings of the workshop on Virtual environments 2002, EGVE '02, pp. 7–12. Eurographics Association, Aire-la-Ville, Switzerland, Switzerland (2002)

177. Hudson, T.C., Lin, M.C., Cohen, J., Gottschalk, S., Manocha, D.: V-collide: Accelerated collision detection for vrml. In: VRML '97: Proceedings of the second symposium on Virtual reality modeling language, pp. 117–ff. ACM, New York, NY, USA (1997)

178. Hui, R., Ouellet, A., Wang, A., Kry, P., Williams, S., Vukovich, G., Peruzzini, W.: Mechanisms for haptic feedback. In: IEEE International Conference on Robotics and Automation, vol. 2, pp. 2138–2143 (1995)

179. Hutchings, B., Grahn, A., Petersen, R.: Multiple-layer cross-field ultrasonic tactile sensor. In: IEEE International Conference on Robotics and Automation, vol. 3, pp. 2522–2528 (1994)

180. Iglesias, R., Casado, S., Gutierrez, T., Barbero, J., Avizzano, C., Marcheschi, S., Bergamasco, M.: Computer graphics access for blind people through a haptic and audio virtual environment. In: The 3rd IEEE International Workshop on Haptic, Audio and Visual Environments and Their Applications, (HAVE 2004). pp. 13–18 (2004)

181. Iglesias, R., Casado, S., Gutierrez, T., Garcia-Alonso, A., Meng Yap, K., Yu, W., Marshall, A.: A peer-to-peer architecture for collaborative haptic assembly. In: IEEE/ACM 10th IEEE International Symposium on Distributed Simulation and Real-Time Applications, (DS-RT'06) pp. 25–34 (2006)

182. Iglesias, R., Prada, E., Uribe, A., Garcia-Alonso, A., Casado, S., Gutierez, T.: Assembly simulation on collaborative haptic virtual environments. In: 15-th International Conference in Central Europe on Computer Graphics, Visualization and Computer Vision, pp. 269–272 (2007)

183. Ignatenko, A., Konushin, A.: A framework for depth image-based modeling and rendering. In: Proceedings of Graphicon-2003, pp. 169–172 (2003)

184. Ikits, M., Brederson, J.D., Hansen, C.D., Johnson, C.R.: A constraint-based technique for haptic volume exploration. In: Proceedings of IEEE Visualization Conference, pp. 263–269 (2003)

185. Immersion: The motiv haptic development platform (2011). URL http://www.immersion. com/products/motiv/index.html

186. Immersion: Touchsense system for mobile devices (2011). URL http://www.immersion.com/ docs/TS-mobile-phones_may10-v2.pdf

187. Ino, S., Shimizu, S., Odagawa, T., Sato, M., Takahashi, M., Izumi, T., Ifukube, T.: A tactile display for presenting quality of materials by changing the temperature of skin surface. In: 2nd IEEE International Workshop on Robot and Human Communication, pp. 220–224 (1993)

188. Ishii, M., Sato, M.: A 3d spatial interface device using tensed strings. Presence: Teleoperators and Virtual Environments 3(1), 81–86 (1994)

189. Jack, D., Boian, R., Merians, A., Tremaine, M., Burdea, G., Adamovich, S., Recce, M., Poizner, H.: Virtual reality-enhanced stroke rehabilitation. IEEE Trans. Neurol. Syst. Rehabil. Eng. 9, 308–318 (2001)

190. Jacko, J.A., Scott, I., Sainfort, F., Moloney, K., Kongnakorn, T., Zorich, B., Emery, V.K.: Effects of multimodal feedback on the performance of older adults with normal and impaired vision. Lect. Notes Comput. Sci. 2615, 3–22 (2003)

191. Jain, R.: Quality of experience. IEEE Multimed. 11(1), 96–95 (2004)

192. Jansson, G., Petrie, H., Colwell, C., Kornbrot, D., Fanger, J., Konig, H., Billberger, K., Hardwick, A., Furner, S.: Haptic virtual environments for blind people: Exploratory experiments with two devices. Int. J. Robot. Res. National J. Virtual Reality 3(4), 8–16 (1998)

193. Johansson, R., Valbo, A.: Tactile sensory coding in the glabrous skin of the human hand. Trends Neurosci. 6(1), 27–32 (1983)

194. Johnson, D., Willemsen, P., Cohen, E.: Six degree-of-freedom haptic rendering using spatialized normal cone search. IEEE Transactions on Visualization and Computer Graphics, 11(6), 661–670 (2005)

195. Johnson, D.E., Cohen, E.: Spatialized normal come hierarchies. In: Proceedings of the 2001 symposium on Interactive 3D graphics, I3D '01, pp. 129–134. ACM, New York, NY, USA (2001)

196. Jones, L.A., Berris, M.: The psychophysics of temperature perception and thermal-interface design. In: S. on Haptic Interfaces for Virtual Environment, T. Systems'2002 (eds.) Symposium on Haptic Interfaces for Virtual Environment and Teleoperator Systems'2002, pp. 137–142 (2002)

197. Jones, M., Marsden, G.: Mobile Interaction Design. John Wiley & Sons Ltd (2006)

198. Kahol, K., Tripathi, P., McDaniel, T., Bratton, L., Panchanathan, S.: Modeling context in haptic perception, rendering, and visualization. ACM Trans. Multimed. Comput. Comm. Appl. 2(3), 219–240 (2006)

199. Kammerl, J., Steinbach, E.: Deadband-based offline-coding of haptic media. In: Proceeding of the 16th ACM international conference on Multimedia, MM '08, pp. 549–558. ACM, New York, NY, USA (2008)

200. Kammerl, J., Vittorias, I., Nitsch, V., Faerber, B., Steinbach, E., Hirche, S.: Perception-based data reduction for haptic force-feedback signals using velocity-adaptive deadbands. Presence (Camb.) 19(5), 450–462 (2010)

201. Katz, D.: The world of touch. Lawrence Erlbaum Associates Inc (1925/1989)
202. Kazerooni, H.: The extender technology: An example of human-machine interaction via the transfer of power and information signals. In: The 4th International Symposium on Experimental Robotics IV, pp. 737–742. Tokyo, Japan (1995)
203. Kim, L., Sukhatme, G.S., Desbrun, M.: A haptic-rendering technique based on hybrid surface representation. IEEE Tran. Comput. Graph. Appl. **24**(2), 66–75 (2004)
204. Kim, S., Hong, D., Hwang, J.H., Kim, B., Choi, S.: Development of an integrated torque sensor-motor module for haptic feedback in teleoperated robot-assisted surgery. In: Technologies for Practical Robot Applications, 2009. TePRA 2009. IEEE International Conference on, pp. 10–15 (2009)
205. Kim, T.H., Ha, J.M., Cho, J.W., You, Y.C., Sung, G.T.: Assessment of the laparoscopic training validity of a virtual reality simulator (lap mentor^{TM}. Korean J. Urol. **50**(10), 989–995 (2010)
206. Kim, Y., Cha, J., Ryu, J., Oakley, I.: A tactile glove design and authoring system for immersive multimedia. IEEE MultiMedia **17**(3), 34–44 (2010)
207. Kim, Y., Kim, S., Ha, T., Oakley, I., Woo, W., Ryu, J.: Air-jet button effects in ar. In: Pan, Z., Cheok, A., Haller, M., Lau, R., Saito, H., Liang, R. (eds.) Advances in Artificial Reality and Tele-Existence, *Lecture Notes in Computer Science*, vol. 4282, pp. 384–391. Springer, Berlin (2006)
208. Kim, Y., Lin, M., Manocha, D.: Deep: Dual-space expansion for estimating penetration depth between convex polytopes. In: Robotics and Automation, 2002. Proceedings. ICRA '02. IEEE International Conference, vol. 1, pp. 921–926 (2002)
209. Kirkpatrick, A., Douglas, S.: Application-based evaluation of haptic interfaces. In: Haptic Interfaces for Virtual Environment and Teleoperator Systems, 2002. HAPTICS 2002. Proceedings. 10th Symposium on, pp. 32–39 (2002)
210. Klatzky, R.L., Lederman, S.J., Reed, C.: There's more to touch than meets the eye: The salience of object attributes for haptics with and without vision. J. Exp. Psychol. Appl. **116**(4), 356–369 (1987)
211. Klatzky, R.L., Loomis, J., Golledge, R., Cicinelli, J., Doherty, S., Pellegrino, J.: Acquisition of route and survey knowledge in the absence of vision. J. Mot. Behav. **22**(1), 19–43 (1990)
212. Kontarinis, D., Howe, R.: Tactile display of vibratory information in tele-operation and virtual environments. Presence (Camb.) **4**(4), 387–402 (1995)
213. Kretz, A., Huber, R., Fjeld, M.: Force feedback slider (ffs): interactive device for learning system dynamics. In: Proceedings of the Fifth IEEE International Conference on Advanced Learning Technologies. Kaohsiung, Taiwan (2005)
214. Krishna, G., Rajanna, K.: Tactile sensor based on piezoelectric resonance. IEEE Sens. J. **4**(5), 691–697 (2004)
215. Kron, A., Schmidt, G.: Bimanual haptic telepresence technology employed to demining operations. In: Proceedings of EuroHaptics 2004. Munich Germany (2004)
216. Kuschel, M., Kremer, P., Hirche, S., Buss, M.: Lossy data reduction methods for haptic telepresence systems. In: Robotics and Automation, 2006. ICRA 2006. Proceedings 2006 IEEE International Conference, pp. 2933–2938 (2006)
217. Lawrence, D., Pao, L., Dougherty, D., Pavlou, Y., Brown, S., Wallace, S.: Human perception of friction in haptic interfaces. In: Proceedings of the Symposium on haptic interfaces for virtual environment and Teleoperator systems, ASME Int. Mech. Eng. Congress and Expo, pp. 287–294 (1998)
218. Lawrence, D.A., Lee, C.D., Pao, L.Y., Novoselov, R.Y.: Shock and vortex visualization using a combined visual/haptic interface. In: Proceedings of IEEE Visualization Conference, pp. 131–137 (2000)
219. Lederman, R., Klatzky, R.L.: Hand movements: A window into haptic object recognition. Cognit. Psychol. **19**(3), 342–368 (1987)
220. Lederman, S., Klatzky, R.: Feeling through a probe. In: D. Systems, C.D.H.I. for Virtual Environments, T. Systems (eds.) Proceedings of the ASME International Mechanical Engineering Congress, vol. 64, pp. 127–131 (1998)

221. Lederman, S., Taylor, M.: Fingertip force, surface geometry and the perception of roughness by active touch. Percept. Psychophys. **12**(5), 401–408 (1972)
222. Lee, H., Takahashi, Y., Miyoshi, T., Inoue, T., Ito, Y., Ikeda Y.and Suzuki, K., Komeda, T.: Basic experiments of upper limb rehabilitation using haptic device system. In: Proceedings of the IEEE 9th International Conference on Rehabilitation Robotics. Chicago, IL, USA (2005)
223. Lemmens, P., Crompvoets, F., Brokken, D., van den Eerenbeemd, J., de Vries, G.J.: A body-conforming tactile jacket to enrich movie viewing. In: EuroHaptics conference, 2009 and Symposium on Haptic Interfaces for Virtual Environment and Teleoperator Systems. World Haptics 2009. Third Joint, pp. 7–12 (2009)
224. Levesque, V.: Blindness technology and haptics. Technical report tr-cim-05.08, Center for Intelligent Machines, Mc-Gill University, Montreal, Quebec, Canada (2005)
225. Levkovich-Maslyuk, L., Ignatenko, A., Zhirkov, A., Konushin, A., Park, I.K., Han, M., Bayakovski, Y.: Depth image-based representation and compression for static and animated 3-D objects. IEEE Trans. Circ. Syst. Video Tech. **14**(7), 1032–1045 (2004)
226. Lin, M., Canny, J.: A fast algorithm for incremental distance calculation. In: Robotics and Automation, 1991. Proceedings., 1991 IEEE International Conference, vol. 2, pp. 1008–1014 (1991)
227. Lin, M.C., Otaduy, M.: Haptic Rendering: Foundations, Algorithms and Applications. A. K. Peters, Ltd., Natick, MA, USA (2008)
228. van der Linde, R., Lammertse, P.: The hapticmaster, a new high performance haptic interface. In: EuroHaptics 2002 (2002)
229. van der Linde, R., Lammertse, P.: Hapticmaster: A generic force controlled robot for human interaction. Ind. Rob. **30**(6), 515–524 (2003)
230. Lindemann, R., Tesar, D.: Construction and demonstration of a 9-string 6 dof force reflecting joystick for telerobotics. In: Proceedings of NASA International Conference on Space Telerobotics, vol. 4, pp. 55–63 (1989)
231. Liu, A., Bhasin, Y., Bower, M.: A haptic-enabled simulator for cricothyroidotomy. Stud. Health. Technol. Inform. **111**, 308–313 (2005)
232. Liu, A., Kaufmann, C., Tanaka, D.: An architecture for simulating needle-based surgical procedures. In: Proceedings of the 4th International Conference on Medical Image Computing and Computer-Assisted Intervention, MICCAI '01, pp. 1137–1144. Springer, London, UK (2001)
233. Liu, A., Tendick, F., Cleary, K., Kaufmann, C.: A survey of surgical simulation: Applications, technology, and education. Presence (Camb.) **12**(6), 599–614 (2003)
234. LLab: Second life. Tech. rep., Second Life (2010). URL http://lindenlab.com/
235. Love, L., Lind, R., Jansen, J.: Mesofluidic actuation for articulated finger and hand prosthetics. In: Intelligent Robots and Systems, 2009. IROS 2009. IEEE/RSJ International Conference, pp. 2586–2591 (2009)
236. M. A. Srinivasan G. L. Beauregard, D.L.B.: The impact of visual information on the haptic perception of stiffness in virtual environments. In: ASME Dynamics Systems and Control Division, pp. 555–559. Atlanta, Georgia, USA (1996)
237. MaClean, K., Roderick, J.: Smart tangible displays in the everyday world: A haptic door knob. In: Advanced Intelligent Mechatronics, 1999. Proceedings. 1999 IEEE/ASME International Conference on, pp. 203–208 (1999)
238. Maekawa, H., Tanie, K., Komoriya, K., Kaneko, M., Horiguchi, C., Sugawara, T.: Development of a finger-shaped tactile sensor and its evaluation by active touch. In: Robotics and Automation, 1992. Proceedings., 1992 IEEE International Conference, vol. 2, pp. 1327–1334 (1992)
239. Magnenat-Thalmann, N., Volino, P., Bonanni, U., Summers, I., Bergamasco, M., Salsedo, F., Wolter, F.: From physics-based simulation to the touching of textiles: The haptex project. Intl. J. Virtual Real. **6**(3), 35–44 (2007)
240. Maingreaud, F., Pissaloux, E., Velazquez, R., Gaunet, F., Hafez, M., Alexandre, J.M.: A dynamic tactile map as a tool for space organization perception: Application to the design of an electronic travel aid for visually impaired and blind people. In: Engineering in Medicine

and Biology Society, 2005. IEEE-EMBS 2005. 27th Annual International Conference of the, pp. 6912–6915 (2005)

241. Mali, U., Munih, M.: Hife-haptic interface for finger exercise. IEEE ASME Trans. Mechatron. **11**, 93–102 (2006)

242. Mansour, M., Eid, M., El Saddik, A.: A multimedia handwriting learning and evaluation tool. In: Proceedings of Intelligent Interactive learning Object Repositories (2007)

243. Marsh, J., Glencross, M., Pettifer, S., Hubbold, R.: A network architecture supporting consistent rich behavior in collaborative interactive applications. Vis. Comput. Graph. IEEE Trans. **12**(3), 405–416 (2006)

244. Marti, G., Rouiller, P., Grange, S., Baur, C.: Biopsy navigator: A smart haptic interface for interventional radiological gestures. In: Proceedings of Computer Assisted Radiology and Surgery (CARS), vol. 1256, pp. 788–793 (2003)

245. Martin, S., Hillier, N.: Characterisation of the novint falocn haptic device for application as a robot manipulator. In: Australasian Conference on Robotics and Automation (ACRA), pp. 291–292 Sydney, Australia (2009)

246. Massie, T., Salisbury, K.: The phantom haptic interface: A device for probing virtual objects. In: ASME Annual Meeting: Symposium on Haptic Interfaces for Virtual Environment and Teleoperator Systems, pp. 295–300 (1994)

247. Mathiowitz, V., Volland, G., Kashman, N., Weber, K.: Adult norms for the box and blocks test of manual dexterity. Am. J. Occup. Ther. **39**(6), 386–391 (1985)

248. Mauve, M., Hilt, V., Kuhmunch, C., Effelsberg, W.: Rtp/i-toward a common application level protocol for distributed interactive media. Multimedia IEEE Trans. **3**(1), 152–161 (2001)

249. McDonnell, K.T., Qin, H., Wlodarczyk, R.A.: Virtual clay: A real-time sculpting system with haptic toolkits. In: Proceedings of the 2001 symposium on Interactive 3D graphics, I3D '01, pp. 179–190. ACM, New York, NY, USA (2001)

250. McLaughlin, M., Rizzo, A., Jung, Y., Peng, W., Yeh, S., Zhu, W.: Haptics-enhanced virtual environments for stroke rehabilitation. In: T.U.C. for Interdisciplinary Research (ed.) The Proceedings of the IPSI (2005)

251. McLaughlin, M., Zimmermann, R., Liu, L., Jung, Y., Peng, W., Jin, S., Stewart, J., Yeh, S.: Integrated voice and haptic support for tele-rehabilitation. In: Fourth Annual IEEE International Conference on Pervasive Computing and Communications Workshops (PERCOMW'06), pp. 590–595 (2006)

252. McNeely, W.A., Puterbaugh, K.D., Troy, J.J.: Six degree-of-freedom haptic rendering using voxel sampling. In: Proceedings of ACM SIGGRAPH, pp. 401–408 (1999)

253. van der Meijden, O.A.J., Schijven, M.P.: The value of haptic feedback in conventional and robot-assisted minimal invasive surgery and virtual reality training: A current review. Surg. Endosc. J. **23**, 1180–1190 (2009)

254. Melder, N., Harwin, W.: Extending the friction cone algorithm for arbitrary polygon based haptic objects. In: Haptic Interfaces for Virtual Environment and Teleoperator Systems, 2004. HAPTICS '04. Proceedings. 12th International Symposium on, pp. 234–241 (2004)

255. Merians, A., Poizner, H., Boian, R., Burdea, G., Adamovich, S.: Sensorimotor training in a virtual reality environment: Does it improve functional recovery poststroke? Neurorehabil. Neural Repair **20**(2), 252–267 (2006)

256. Michitaka, H., Koichi, H., Tetsuro, O., Makoto, S., Mutsuhiro, N.: Construction of haptic server for remote cooperative work. In: Proceedings of the Virtual Reality Society of Japan Annual Conference, pp. 269–272 (2000)

257. Microsoft: Microsoft directinput overview (2011). URL MicrosoftDirectInput:http://msdn. microsoft.com/en-us/library/ms810471.aspx

258. Miner, N., Gillespie, B., Caudell, T.: Examining the influence of audio and visual stimuli on a haptic display. In: Proceedings of the 1996 IMAGE Conference (1996)

259. Minsky, M., Ming, O.Y., Steele, O., Brooks, F.P. Jr., Behensky, M.: Feeling and seeing: Issues in force display. SIGGRAPH Comput. Graph. **24**(2), 235–241 (1990)

260. Minsky, M.D.R.: Computational haptics: The sandpaper system for synthesizing texture for a force-feedback display. Ph.D. thesis, Massachusetts Institute of Technology (1995)
261. Monkman, G.J.: 3-d tactile image display. Sens. Rev. **13**(2), 27–31 (1993)
262. Moore, D.: The importance of touch. http://library.adoption.com/articles/the-importance-of-touch.html
263. Morris, D., Neel, J., Salisbury, K.: Haptic battle pong: High-degree-of-freedom haptics in a multiplayer gaming environment. In: Experimental Gameplay Workshop, GDC (2004)
264. Nicholls, H., Lee, M.: A survey of robot tactile sensing technology. Int. J. Rob. Res. **8**, 3–30 (1989)
265. Nilsen, T., Linton, S., Looser, J.: Motivations for ar gaming. In: Proceedings Fuse, New Zealand Game Developers Conference, pp. 86–93. Dunedin, New Zealand (2004)
266. Nilsson, D., Aamisepp, H.: Haptic hardware support in a 3d game engine. Master's thesis, Department of Computer Science, Lund University (2003)
267. Novint: Novint Falcon SDK. Novint Technologies Inc (NVNT), 4601 Paradise Blvd. NW Albuquerque, NM 87114 (2010)
268. Oakley, I., Brewster, S., Gray, P.: Can you feel the force? an investigation of haptic collaboration in shared editors. In: Proceedings of Eurohaptics, pp. 54–59 (2001)
269. Okamura, A.: Haptic feedback in robot-assisted minimally invasive surgery. Curr. Opin. Urol. **19**(1), 102–107 (2009)
270. O'Malley, M., Goldfarb, M.: The effect of force saturation on the haptic perception of detail. Mechatron. IEEE/ASME Trans. **7**(3), 280–288 (2002)
271. O'Modhrain, S., Oakley, I.: Touch tv: Adding feeling to broadcast media. In: Proceedings of the European Conference on Interactive Television: from Viewers to Actors (2003)
272. O'Modhrain, S., Oakley, I.: Adding interactivity: Active touch in broadcast media. In: Haptic Interfaces for Virtual Environment and Teleoperator Systems, 2004. HAPTICS '04. Proceedings. 12th International Symposium on, pp. 293–294 (2004)
273. Orozco, M., Asfaw, Y., Adler, A., Shirmohammadi, S., El Saddik, A.: Automatic identification of participants in haptic systems. In: Proceedings of the IEEE Instrumentation and Measurement Technology Conference (IEEE IMTC '05). Ottawa, ON, Canada, (2005)
274. Orozco, M., Asfaw, Y., Shirmohammadi, S., Adler, A., El Saddik, A.: Haptic-based biometrics: A feasibility study. In: IEEE Virtual Reality Conference. Alexandria, Virginia, USA (2006)
275. Orozco, M., Malek, B., Eid, M., El Saddik, A.: Haptic-based sensible graphical password. In: Proceedings of Virtual Concept 2006. Playa Del Carmen, Mexico, (2006)
276. Otaduy, M.A., Lin, M.C.: Sensation preserving simplification for haptic rendering. In: ACM SIGGRAPH 2003 Papers, SIGGRAPH '03, pp. 543–553. ACM, New York, NY, USA (2003)
277. Ouhyoung, M., Tsai, W., Tsai, M., Wu, J., Huang, C., Yang, T.: A low-cost force feedback joystick and its use in pc video games. IEEE Trans. Consum. Electron. **41**(3), 787–794 (1995)
278. Paillard, J.: Motor control, today and tomorrow. In: Body Schema and Body Image – A Double Dissociation in Deafferented Patient, pp. 197–214. Academic publishing House, Sophia (1999)
279. Pang, X.D., Tan, H.Z., Durlach, N.I.: Manual discrimination of force using active finger motion. Percept. Psychophys. **49**(6), 531–540 (1991)
280. Paranjape, R.P., Johnson, M.J., Ramachandran, B.: Assessing impaired arm use and learned bias after stroke using unimanual and bimanual steering tasks. Engineering in Medicine and Biology Society, 2006. EMBS '06. 28th Annual International Conference of the IEEE, pp. 3958–3961, 30 Aug 2006 to 3 Sept 2006 doi: 10.1109/IEMBS.2006.260026 URL: http://ieeexplore.ieee.org/stamp/stamp.jsp?tp=&arnumber=4462666&isnumber=4461641
281. Park, K.S., Kenyon, R.: Effects of network characteristics on human performance in a collaborative virtual environment. In: Virtual Reality, 1999. Proceedings., IEEE, pp. 104–111 (1999)
282. Parton, A., Malhotra, P., Husain, M.: Hemispatial neglect. J. Neurol. Neurosurg. Psychiatry **75**(1), 13–21 (2004)

283. Pascale, M., Mulatto, S., Prattichizzo, D.: Bringing haptics to second life for visually impaired people. In: Proceedings of the 6th international conference on Haptics: Perception, Devices and Scenarios, EuroHaptics '08, pp. 896–905. Springer, Berlin (2008)

284. Pawlak, A.M.: Sensors and actuators in mechatronics: Design and applications. Taylor and Francis, Connecticut, USA (2007)

285. Payeur, P., Pasca, C., Cretu, A.M., Petriu, E.: Intelligent haptic sensor system for robotic manipulation. Instrum. Meas. IEEE Trans. 54(4), 1583–1592 (2005)

286. Peshkin, M., Colgate, J., Moore, C.: Passive robots and haptic displays based on nonholonomic elements. In: Robotics and Automation, 1996. Proceedings., 1996 IEEE International Conference on, vol. 1, pp. 551–556 (1996)

287. Petriu, E., Yeung, S., Das, S., Cretu, A.M., Spoelder, H.: Robotic tactile recognition of pseudorandom encoded objects. Instrum. Meas., IEEE Trans. 53(5), 1425–1432 (2004)

288. Picinbono, G., Lombardo, J., Delingette, H., Ayache, N.: Improving realism of a surgery simulator: Linear anisotropic elasticity, complex interactions and force extrapolation. J. Vis. Comput. Animation 13(3), 147–167 (2002)

289. Ping, L., Wenjuan, L., Zengqi, S.: Transport layer protocol reconfiguration for network-based robot control system. In: Networking, Sensing and Control, 2005. Proceedings. 2005 IEEE, pp. 1049–1053 (2005)

290. Pohja, S.: Survey of studies on tactile senses. Tech. rep., European Research Consortium for Informatics and Mathematics at SICS (2001)

291. Polushin, I., Liu, P., Lung, C.H.: A control scheme for stable force-reflecting teleoperation over ip networks. In: 2005 IEEE/RSJ International Conference on Intelligent Robots and Systems, 2005, pp. 2731–2736 (2005)

292. Polushin, I., Liu, P., Lung, C.H.: Force reflection algorithm for improved transparency in bilateral teleoperation with communication delay. In: Proceedings 2006 IEEE International Conference on Robotics and Automation, 2006, pp. 2914–2920 (2006)

293. Popescu, V., Burdea, G., Bouzit, M., Hentz, V.: A virtual-reality based telerehabilitation system with force feedback. IEEE Trans. Inform. Technol. Biomed. 4(1), 45–51 (2000)

294. Portillo, O., Avizzano, C., Raspolli, M., Bergamasco, M.: Haptic desktop for assisted handwriting and drawing. In: Proceedings of the IEEE International Workshop on Robots and Human Interactive Communication. Nashville, TN (2005)

295. Pressure Profile, I.: http://www.pressureprofile.com. Tech. rep., Pressure Profile, Inc. (2006)

296. Rahman, A., Eid, M., El Saddik, A.: Kissme: Bringing virtual events to the real world. In: Virtual Environments, Human-Computer Interfaces and Measurement Systems, VECIMS 2008, pp. 102–105. Istanbul, Turkey (2008)

297. Rahman, A.M., Hossain, S., El Saddik, A.: Bridging the gap between virtual and real world by bringing an interpersonal haptic communication system in second life. In: IEEE International Symposium on Multimedia (ISM2010). Taiwan (2010)

298. Raibert, M.H., Tanner, J.E.: Design and implementation of a vlsi tactile sensing computer. Int. J. Rob. Res. 1(3), 3–18 (1982)

299. Ramsey, A.: Investigation of physiological measures relative to self-report of virtual reality induced sickness and effects (vrise). In: The International Workshop on Motion Sickness: Medical and Human Factors (1997)

300. Ramstein, C., Hayward, V.: The pantograph: A large workspace haptic device for multimodal human computer interaction. In: Conference companion on Human factors in computing systems, CHI '94, pp. 57–58. ACM, New York, NY, USA (1994)

301. Raymaekers, C., De Boeck, J., Coninx, K.: An empirical approach for the evaluation of haptic algorithms. In: Eurohaptics Conference, 2005 and Symposium on Haptic Interfaces for Virtual Environment and Teleoperator Systems, 2005. World Haptics 2005. First Joint, pp. 567–568 (2005)

302. Reiley, C., Akinbiyi, T., Burschka, D., Chang, D., Okamura, A., Yuh, D.: Effects of visual force feedback on robot-assisted surgical task performance. Thorac. Cardiovasc. Surg. 135, 196–202 (2008)

303. Reimer, E., Baldwin, L.: Cavity sensor technology for low cost automotive safety & control devices. In: Proceedings of the Air Bag Technology'99 (1999)
304. Rejin, I., Stanojevic, M., Reljin, B.: Modified round-robin scheduler for pareto traffic streams. In: Telecommunications in Modern Satellite, Cable and Broadcasting Service, 2001. TELSIKS 2001. 5th International Conference on, vol. 1, pp. 25 –28 vol.1 (2001)
305. Révész, G.: Psychology and Art of the Blind. Longmans, Green and Co. (1950). (Translated by H. A. Wolff)
306. Reynaerts, D., Brussel, H.V.: Tactile sensing data interpretation for object manipulation. Sens. Actuators A Phys. **37–38**, 268–273 (1993). (Proceedings of Eurosensors VI)
307. Riener, R., Frey, M., Proll, T., Regenfelder, F., Burgkart, R.: Phantom-based multimodal interactions for medical education and training: The munich knee joint simulator. IEEE Trans. Inform. Technol. Biomed. **8**, 208–216 (2004)
308. Roberts, D., Wolff, R.: Controlling consistency within collaborative virtual environments. In: Distributed Simulation and Real-Time Applications, 2004. DS-RT 2004. Eighth IEEE International Symposium on, pp. 46–52 (2004)
309. Robles-De-La-Torre, G.: The importance of the sense of touch in virtual and real environments. IEEE Multimedia J **13**(3), 24–30 (2006) (Special issue on Haptic User Interfaces for Multimedia Systems)
310. Robles-De-La-Torre, G., Hayward, V.: Force can overcome object geometry in the perception of shape through active touch. Nature **412**, 445–448 (2001)
311. Rohling, R., Hollerbach, J., Jacobsen, S.: Optimized fingertip mapping: A general algorithm for robotic hand teleoperation. Presence Teleoperators Virtual Environ. **2**(3), 203–220 (1993)
312. Rondao Alface, P., Macq, B.: Shape quality measurement for 3D watermarking schemes. In: Delp, E.J. III , Wong, P.W. (eds.) Society of Photo-Optical Instrumentation Engineers (SPIE) Conference Series. Society of Photo-Optical Instrumentation Engineers (SPIE) Conference Series, vol. 6072, pp. 622–634 (2006)
313. Rosenberg, L.: Using force feedback to enhance human performance in graphical user interfaces. In: ACM CHI 96, pp. 291–292. Vancouver, British Columbia, Canada (1996)
314. Rosenberg, L., Adelstein, B.: Perceptual decomposition of virtual haptic surfaces. In: Virtual Reality, 1993. Proceedings., IEEE 1993 Symposium on Research Frontiers in, pp. 46–53 (1993)
315. Rosenberg, L.B., Chang, D.C.: Method and apparatus for providing dynamic force sensations for force feedback computer applications (2006). URL http://patent.ipexl.com/US/7039866.html
316. Ross, H.E., Murray, D.J.: E.H. Weber on the Tactile Senses. Tayor & Francis, Hove, UK (1996)
317. Rothchild, R.A., Mann, R.W.: An emg-controlled force sensing proportional rate elbow prosthesis. In: Proceeding of the 1966 Symposium on Biomedical Engineering, pp. 106–109. Milwaukee, Wisconsin (1966)
318. Rovers, A., van Essen, H.: Him: A framework for haptic instant messaging. In: CHI '04 extended abstracts on Human factors in computing systems, CHI EA '04, pp. 1313–1316. ACM, New York, NY, USA (2004)
319. Ruffaldi, E., Morris, D., Edmunds, T., Barbagli, F., Pai, D.: Standardized evaluation of haptic rendering systems. In: Haptic Interfaces for Virtual Environment and Teleoperator Systems, 2006 14th Symposium, pp. 225–232. (2006)
320. Ruspini, D.C., Kolarov, K., Khatib, O.: The haptic display of complex graphical environments. In: SIGGRAPH '97: Proceedings of the 24th annual conference on Computer graphics and interactive techniques, pp. 345–352. ACM Press/Addison-Wesley Publishing Co., New York, NY, USA (1997)
321. Russell, R.: Compliant-skin tactile sensor. In: Robotics and Automation. Proceedings. 1987 IEEE International Conference on, vol. 4, pp. 1645–1648 (1987)
322. Ryu, J., Choi, S.: Posvibeditor: Graphical authoring tool of vibrotactile patterns. In: IEEE International Workshop Haptic Audio, and Visual Environments, and Games, pp. 120–125. (2008)

323. Sakr, N., Georganas, N., Zhao, J.: Perceptibility of watermarked 3d meshes using a multimodal visuo-haptic interface. In: Virtual Environments, Human-Computer Interfaces and Measurement Systems, 2008. VECIMS 2008. IEEE Conference on, pp. 20–24 (2008)
324. Sakr, N., Georganas, N., Zhao, J., Petriu, E.: Perceptibility of digital watermarks embedded in 3d meshes using unimodal and bimodal hapto-visual interaction. In: Instrumentation and Measurement Technology Conference, 2009. I2MTC '09. IEEE, pp. 987–991 (2009)
325. Sakr, N., Georganas, N., Zhao, J., Petriu, E.: Multimodal vision-haptic perception of digital watermarks embedded in 3-d meshes. Instrum. Meas. IEEE Trans. 59(5), 1047–1055 (2010)
326. Salcudean, S.E., Parker, N.: 6-dof desk-top voice-coil joystick. In: Proceedings of the 6th Annual Symposium on Haptic Interfaces for Virtual Environments and Teleoperation Systems. Dallas, Texas (1997)
327. Saleh, M., Habib, I., Saadawi, T.: Simulation analysis of a communication link with statistically multiplexed bursty noise sources. Sel. Areas Commun. IEEE J. 11(3), 432–442 (1993)
328. Salisbury, J.K., Tarr, C.: Haptic rendering of surfaces defined by implicit functions. In: Proceedings of Symp. Haptic Interfaces for Virtual Environment and Teleoperator Systems, pp. 61–68 (1997)
329. Salisbury, K., Conti, F., Barbagli, F.: Haptic rendering: Introductory concepts. IEEE Comput. Graph. Appl. 24(2), 24–32 (2004)
330. Sathian, K.: Tactile sensing of surface features. Trends Neurosci. 12, 513–519 (1989)
331. Sato, M., Hirata, Y., Kawarada, H.: Space interface device for artificial reality – spidar. J. Rob. Mechatron. 9(3), 177–184 (1997)
332. van Scoy, F., Kawai, T., Darrah, M., Rash, C.: Haptic display of mathematical functions for teaching mathematics to students with vision disabilities: Design and proof of concept. In: Haptic Human-Computer Interaction: First International Workshop. Glasgow, UK. (2000)
333. van Scoy, F., Kawai, T., Fullmer, A., Stamper, K., Wojciechowksa, I., Perez, A., Vargas, J., Martinex, S.: The sound and touch of mathematics: A prototype system. In: Proceedings of the PHANToM Users Group PUG. Aspen, CO. (2001)
334. Sensable: Openhaptics toolkit: http://www.sensable.com/products-openhaptics-toolkit.htm (2011). URL http://www.sensable.com/products-openhaptics-toolkit.htm
335. Sensable: Phantom omni haptic device (2011). URL http://www.sensable.com/haptic-phantom-omni.htm
336. Seo, Y., Lee, B., Kim, Y., Kim, J., Ryu, J.: K-hapticmodeler^{TM}: A haptic modeling scope and basic framework. In: IEEE International Workshop Haptic Audio, and Visual Environments, and their Application, pp. 136–141 (2007)
337. Shahabi, C., Ortega, A., Kolahdouzan, M.: A comparison of different haptic compression techniques. In: Multimedia and Expo, 2002. ICME '02. Proceedings. 2002 IEEE International Conference on, vol. 1, pp. 657–660 vol.1 (2002)
338. Shakra, I., Orozco, M., El Saddik, A., Shirmohammadi, S., Lemaire, E.: Haptic instrumentation for physical rehabilitation of stroke patients. In: Proceedings of the 2006 IEEE International Workshop on Medical Measurement and Applications. Benevento, Italy (2006)
339. Shakra, I., Orozco, M., El Saddik, A., Shirmohammadi, S., Lemaire, E.: Vr-based hand rehabilitation using a haptic-based framework. In: Proceedings of the 23rd IEEE Instrumentation and Measurement Technology Conference, pp. 1178–1181. Sorrento, Italy (2006)
340. Shen, X., Bogsanyi, F., Ni, L., Georganas, N.: A heterogeneous scalable architecture for collaborative haptics environments. In: Haptic, Audio and Visual Environments and Their Applications, 2003. HAVE 2003. Proceedings. The 2nd IEEE Internatioal Workshop on, pp. 113–118 (2003)
341. Shen, X., Zhou, J., El Saddik, A., Georganas, N.: Architecture and evaluation of tele-haptic environments. In: Distributed Simulation and Real-Time Applications, 2004. DS-RT 2004. Eighth IEEE International Symposium on, pp. 53–60 (2004)
342. Shen, X., Zhou, J., Georganas, N.D.: Evaluation patterns of tele-haptics. In: Electrical and Computer Engineering, 2006. CCECE '06. Canadian Conference on, pp. 1542–1545 (2006)
343. Shepherd, G.M.: Neurobiology. New York: Oxford University Press (1988)

344. Shields, B., Main, J., Peterson, S., Strauss, A.: An anthropomorphic hand exoskeleton to prevent astronaut hand fatigue during extravehicular activities. Syst. Man Cybern. Part A Syst. Hum. IEEE Trans. **27**(5), 668–673 (1997)

345. Shimojo, M., Ishikawa, M., Kanaya, K.: A flexible high resolution tactile imager with video signal output. In: Robotics and Automation, 1991. Proceedings., 1991 IEEE International Conference on, pp. 384–389 vol. 1 (1991)

346. Shin, H., Lee, J., Park, J., Kim, Y., Oh, H., Lee, T.: A tactile emotional interface for instant messenger chat. In: Smith, M., Salvendy, G. (eds.) Human Interface and the Management of Information. Interacting in Information Environments. Lecture Notes in Computer Science, vol. 4558, pp. 166–175. Springer, Berlin/Heidelberg (2007)

347. Shirmohammadi, S., Georganas, N.D.: An end-to-end communication architecture for collaborative virtual environments. Comput. Netw. **35**, 2–3 (2001)

348. Siegel, D., Drucker, S., Garabieta, I.: Performance analysis of a tactile sensor. In: Robotics and Automation. Proceedings. 1987 IEEE International Conference on, vol. 4, pp. 1493–1499 (1987)

349. Siira, J., Pai, D.: Haptic texturing-a stochastic approach. In: Robotics and Automation, 1996. Proceedings., 1996 IEEE International Conference on, vol. 1, pp. 557–562 (1996)

350. de Silva, C.W.: Sensors and Actuators: Control Systems Instrumentation. CRC, Connecticut, USA (2007)

351. Smith, S., Williams, G.: A visualization of music. In: Visualization '97., Proceedings, pp. 499–503 (1997)

352. Snowglobe: Snowglobe (2010). URL http://wiki.secondlife.com/wiki/Snowglobe

353. Souayed, R., Gaiti, D., Yu, W., Dodds, G., Marshall, A.: Investigation of network issues for distributed haptic virtual environment applications. In: Information and Communication Technologies: From Theory to Applications, 2004. Proceedings. 2004 International Conference on, pp. 677–678 (2004)

354. Souayed, R.T., Gaiti, D., Pujolle, G., Yu, W., Gu, Q., Marshall, A.: Haptic virtual environment performance over ip networks: A case study. In: Distributed Simulation and Real-Time Applications, 2003. Proceedings. Seventh IEEE International Symposium on, pp. 181–189 (2003)

355. Spano, S., Bourbakis, N.: Design and implementation of a low cost multi-fingered robotic hand using a method of blocks. J. Int. Rob. Syst. **30**, 209–226 (2001)

356. Speeter, T.H.: Flexible, piezoresistive touch sensing array. In: Proceedings of the Society of Photo-Optical Instrumentation Engineers, vol. 1005, p. 31 (1988)

357. Srinivasan, M.A., Basdogan, C.: Haptics in virtual environments: Taxonomy, research status, and challenges. Comput. Graph. **21**(4), 393–404 (1997). (Haptic Displays in Virtual Environments and Computer Graphics in Korea)

358. Srinivasan, M.A., LaMotte, R.H.: Tactual discrimination of softness. J. Neurophysiol. **73**(1), 88–101 (1995)

359. Stansfield, S.: Primitives, features, and exploratory procedures: Building a robot tactile perception system. In: Robotics and Automation. Proceedings. 1986 IEEE International Conference on, vol. 3, pp. 1274–1279 (1986)

360. Stevens, J.C.: Thermal sensibility, The Psychology of Touch, M.A. Heller and W. Schiff, eds., Lawrence Erlbaum, pp. 61–90 (1991)

361. van Strijp, C., Langen, H., Onosato, M.: The application of a haptic interface on microassembly. In: Haptic Interfaces for Virtual Envionment and Teleoperator System HAPTIC 06, pp. 289–293 (2006)

362. Sugiyama, S., Kawahata, K., Yoneda, M., Igarashi, I.: Tactile image detection using a 1k-element silicon pressure sensor array. Sens. Actuators A Phys. **22**(1-3), 397–400 (1990)

363. Tan, H., Srinivasan, M., Eberman, B., Cheng, B.: Human factors for the design of force-reflecting haptic interfaces. ASME Dyn. Syst. Control **55-1**, 353–359 (1994)

364. Tan, H.Z., Gray, R., Young, J.J., Traylor, R.: A haptic back display for attentional and directional cueing. Haptics-e **3**(1) (2003)

365. Tapson, J., Gurari, N., Diaz, J., Chicca, E., Sander, D., Pouliquen, P., Etienne-Cummings, R.: The feeling of color: A haptic feedback device for the visually disabled. In: Biomedical Circuits and Systems Conference, 2008. BioCAS 2008. IEEE, pp. 381–384 (2008)

366. Taylor, E., Murray, J.: Medical procedure classification in canada–where are we going? Tech. rep., Canadian Centre for Health Information, Statistics, Canada (1993)

367. Taylor, P., Hosseini-Sianaki, A., Varley, C.: Surface feedback for virtual environment systems using electrorhelological fluids. Int. J. Mod. Phys. B 10(23-24), 3011–3018 (1996)

368. Teh, J., Lee, S.P., Cheok, A.D.: Internet.pajama. In: Proceedings of the 2005 International Conference on Augmented Tele-Existence, ICAT '05, pp. 274–274. ACM, New York, NY, USA (2005)

369. Tholey, G., Desai, J.P., Castellanos, A.E.: Force feedback plays a significant role in minimally invasive surgery: Results and analysis. Ann. Surg. 241(1), 102–109 (2005)

370. Thompson, D.E., Giurintano, D.J.: A kinematic model of the flexor tendons of the hand. J. Biomechan. 22(4), 327–331, 333–334 (1989)

371. Thompson, T.V., Cohen, E.: Direct haptic rendering of complex trimmed NURBS models. In: Proceedings of Symposium Haptic Interfaces for Virtual Environment and Teleoperator Systems, pp. 89–96 (1999)

372. Thompson, T.V., Nelson, D.D., Cohen, E., Hollerbach, J.: Maneuverable NURBS models within a haptic virtual environment. In: Proceedings of Symposium Haptic Interfaces for Virtual Environment and Teleoperator Systems, pp. 37–44 (1997)

373. Tsagarakis, N., Caldwell, D.: A 5 dof haptic interface for pre-operative planning of surgical access in hip arthroplasty. In: First Joint Eurohaptics Conference and Symposium on Haptic Interfaces for Virtual Environment and Teleoperator Systems, pp. 519–520 (2005)

374. Tsumaki, Y., Naruse, H., Nenchev, D., Uchiyama, M.: Design of a compact 6-dof haptic interface. In: Robotics and Automation, 1998. Proceedings. 1998 IEEE International Conference on, vol. 3, pp. 2580–2585 (1998)

375. Turner, M.L., Gomez, D.H., Tremblay, M.R., Cutkosky, M.R.: Preliminary tests of an arm-grounded haptic feedback device in telemanipulation. In: Proceedings of the ASME Dynamic Systems and Control Division, vol. 64, pp. 145–149 (1998)

376. Tyberghein, J., Sunshine, E., Richter, F., Svanfeldt, M., Zabolotny, A., Galbraith, S., Wyett, P., Voase, M.: Crystal space. Http://www.crystalspace3d.org/docs/online/1.4/manual/

377. Tzafestas, C., Coiffet, P.: Computing optimal forces for generalised kinesthetic feedback on the human hand during virtual grasping and manipulation. In: Robotics and Automation, 1997. Proceedings., 1997 IEEE International Conference on, vol. 1, pp. 118–123 (1997)

378. Uchimura, Y., Yakoh, T.: Bilateral robot system on the real-time network structure. Ind. Electron. IEEE Trans. 51(5), 940–946 (2004)

379. Uchiyama, M., Tsumaki, Y., Yoon, W.K.: Design of a compact 6-dof haptic device to use parallel mechanisms. In: International Symposium on Robotics Research. San Francisco, CA (2005)

380. Ueberle, M., Buss, M.: Design, control, and evaluation of a new 6 dof haptic device. In: Intelligent Robots and Systems, 2002. IEEE/RSJ International Conference on, vol. 3, pp. 2949–2954 (2002)

381. Ueberle, M., Mock, N., Buss, M.: Vishard10, a novel hyper-redundant haptic interface. In: Haptic Interfaces for Virtual Environment and Teleoperator Systems, 2004. HAPTICS '04. Proceedings. 12th International Symposium on (2004)

382. Ullrich, C.: Haptic phone gaming study. Tech. rep., Immersion Corporation (2005)

383. Ursino, M., Tasto, J.L., Nguyen, B.H., Cunningham, R., Merril, G.L., CathSim: an intravascular catheterization simulator on a PC. Proc. Medicine Meets Virtual Reality Conf., IOS Press, Amsterdam, pp. 360–366, 1999

384. Vega, C.: Sensurround trademark to enhance the audio experience during film screenings (1970)

385. W3C: Virtual reality modeling language (2011). URL VirtualRealityModelingLanguage: http://www.w3.org/MarkUp/VRML

386. Walker, S.P., Salisbury, J.K.: Large haptic topographic maps: Marsview and the proxy graph algorithm. In: I3D '03: Proceedings of the 2003 symposium on Interactive 3D graphics, pp. 83–92. ACM, New York, NY, USA (2003)
387. Wall, S., Harwin, W.: Interaction of visual and haptic information in simulated environments: Texture perception. In: Proceedings of the 1st Workshop on Haptic Human Computer Interaction, pp. 39–44 (2000)
388. Wang, D., Zhang, Y., Yao, C.: Machine-mediated motor skill training method in haptic-enabled chinese handwriting simulation system. In: Intelligent Robots and Systems, 2006 IEEE/RSJ International Conference on, pp. 5644–5649 (2006)
389. Watt, A., Policarpo, F.: The Computer Image. Addison Wesley, Harlow, UK (1999)
390. Way, T., Barner, K.: Automatic visual to tactile translation. ii. evaluation of the tactile image creation system. Rehabil. Eng. IEEE Trans. 5(1), 95–105 (1997)
391. Wei, C., Marsden, G., Gain, J.: Novel interface for first person shooting games on pdas. In: Press, A. (ed.) Proceedings of the 20th Australasian Conference on Computer-Human Interaction: Designing for Habitus and Habitat OZCHI '08:, vol. 287, pp. 113–121. OZCHI, Cairns, A ustralia (2008)
392. Whalen, T., Noel, S., Stewart, J.: Measuring the human side of virtual reality. In: Virtual Environments, Human-Computer Interfaces and Measurement Systems, 2003. VECIMS '03. 2003 IEEE International Symposium, pp. 8–12. (2003)
393. Whitney, D.E.: Historical perspective and state of the art in robot force control. Int. J. Rob. Res. 6, 3–14 (1987)
394. Wing, A., Haggard, P., Flanagan, J.: Hand and Brain: Neurophysiology and Psychology of Hand Movement. Academic, San Diego (1996)
395. Wirz, R., Ferre, M., Marín, R., Barrio, J., Claver, J.M., Ortego, J.: Efficient transport protocol for networked haptics applications. In: Proceedings of the 6th international conference on Haptics: Perception, Devices and Scenarios, EuroHaptics'08, pp. 3–12. Springer, Berlin (2008)
396. Wong, N., Salcudean, S., Hollis, R.: Design and control of a force-reflecting teleoperation system with magnetically leviated master and wrist. IEEE Trans. Rob. Autom. 11, 844–858 (1995)
397. Wu, W., Arefin, A., Rivas, R., Nahrstedt, K., Sheppard, R., Yang, Z.: Quality of experience in distributed interactive multimedia environments: toward a theoretical framework. In: Proceedings of the 17th ACM international conference on Multimedia, MM '09, pp. 481–490. ACM, New York, NY, USA (2009)
398. Xin, H., Zelek, J., Carnahan, H.: Laparoscopic surgery, perceptual limitations and force: A review. In: First Canadian Student Conference on Biomedical Computing, vol. 144. Kingston, Ontario, Canada (2006)
399. Yamaguchi, T., Akabane, A., Murayama, J., Sato, M.: Automatic generation of haptic effect into published 2d graphics. In: Proceedings of the EuroHaptics (2006)
400. Yanagida, Y., Kakita, M., Lindeman, R., Kume, Y., Tetsutani, N.: Vibrotactile letter reading using a low-resolution tactor array. In: Haptic Interfaces for Virtual Environment and Teleoperator Systems, 2004. HAPTICS '04. Proceedings. 12th International Symposium, pp. 400–406 (2004)
401. Ye, Y., Liu, P.: Improving haptic feedback fidelity in wave-variable-based teleoperation orientated to telemedical applications. IEEE Trans. Instrum. Meas. 58(8), 2847–2855 (2009)
402. Yeung, S., Petriu, E., McMath, W., Petriu, D.: High sampling resolution tactile sensor for object recognition. Instrum. Meas. IEEE Trans. 43(2), 277–282 (1994)
403. Yoshida, H., Noma, H., Tetsutani, N., Kurumisawa, J.: Sumi-nagashi: Adding corporeality and tactile sensation to digital painting. Comput. Graph. Appl. IEEE 24(1), 1 pp. (2004)
404. Yu, W., Pernalete, N., Dubey, R.: Motion therapy for persons with disabilities using hidden markov model based skill learning. In: Proceedings of the IEEE International Conference on Robotics and Automation (ICRA '04). New Orleans, LA (2004)

405. Yu, W., Ramloll, R., Brewster, S.: Haptic graphs for blind computer users. Haptic Human-Computer Interaction, pp. 41–51. Springer, Berlin/Heidelberg (2001)
406. Zadeh, M., Wang, D., Kubica, E.: Perception-based lossy haptic compression considerations for velocity-based interactions. Multimedia Syst. **13**, 275–282 (2008)
407. Zaeh, M., Clarke, S., Hinterseer, P., Steinbach, E.: Telepresence across networks: A combined deadband and prediction approach. In: Information Visualization, 2006. IV 2006. Tenth International Conference, pp. 597–604 (2006)
408. Zajeganovic-Ivancic, M., Reljin, I., Reljin, B.: Video multicoder with neural network control. In: Neural Network Applications in Electrical Engineering, 2008. NEUREL 2008. 9th Symposium, pp. 187–191 (2008)
409. Zhou, J., Shen, X., Georganas, N.: Haptic tele-surgery simulation. In: Proceedings of the 3rd IEEE International Workshop on Haptic, Audio and Visual Environments and Their Applications, pp. 99–104 (2004)
410. Zilles, C., Salisbury, K.: A constraint based god-object method for haptic display. In: IEE/RSJ International Conference on Intelligent Robots and Systems, Human Robot Interaction, and Cooperative Robots, vol. 3, pp. 146–151 (1995)

Index

A. El Saddik et al., *Haptics Technologies*, Springer Series on Touch and Haptic Systems,   217
DOI 10.1007/978-3-642-22658-8, © Springer-Verlag Berlin Heidelberg 2011